BY THE EDITORS OF CONSUMER GUIDE®



# MODEL CARS

**HARPER COLOPHON BOOKS**

Harper & Row, Publishers

New York, Hagerstown, San Francisco, London

# Contents




Printed in the United States of America

Published simultaneously in Canada by Fitzhenry & Whiteside Limited, Toronto, Canada.

Library of Congress Catalog Card Number: 78-3188
ISBN: 0-06-010844-4

First Harper Colophon edition published 1978.

PEOPLE HAD BEEN dreaming the dream for centuries before Karl Benz set the three wheels of his invention on a muddy path in 1885. Leonardo da Vinci had made sketches of self-propelled vehicles 300 years earlier. Outlandish carts designed to be pushed along by the wind had been built, but they went nowhere. Steam had been used to power other primitive horseless carriages. But it was not until Benz developed a car with an internal-combustion engine that the automobile revolution began.

Henry Ford's mass-production techniques enabled many internal-combustion cars to be built each week and kept their price low enough so that many people could afford them. By the early 1900s, the automobile had ceased to be a novelty and had become much more of a daily necessity.

Yet as necessary as it is to today's way of life, the automobile is still an object of affection and a source of excitement. The romance of the motorcar includes the names of Cord, Duesenberg, Stutz and Hudson. Henry Ford, Louis Chevrolet and Ransom E. Olds have earned a place for themselves on the list of American pioneers. Bobby Unser, A. J. Foyt and Mario Andretti are folk heroes—superstars of the sporting world—who use their automobiles to entertain thousands of racing fans.

There are some people who can collect full-size automobiles. They subscribe to magazines that are filled with ads for these high-priced collector's items and travel around the country in search of a particular Rolls-Royce to complete their set. There

1

# Introduction

are other people who have the skill and the money to race, to sit behind the wheel of a fast and agile sports car and feel the surge of adrenalin as they push themselves and their machines harder and harder.

The rest of us may not be able to collect the real thing, but we can build and collect miniature replicas. We may not be able to race in the Indianapolis 500, but we can feel the excitement of racing by running slot cars around a rec room track or by maneuvering a gasoline-powered model around an outdoor course with radio waves. These activities are what *Model Cars* is all about.

The building of model cars from kits or from scratch is a hobby for the craftsman, a person who enjoys the tranquility of careful assembly and the artistry of decoration. Racing models is a hobby for those who enjoy the roar of small but powerful engines and the challenge of competition at speeds of up to 50 miles an hour.

No matter which of the forms of automotive modeling interests you, there is much more to the hobby than simply buying a kit and putting it together or purchasing a racing car and running it around the track. The customizing of static models and the fine-tuning of radio-control and slot-car models allows the hobbyist to put much of his or her personality into the car.

How much you put into the hobby, of course, is up to you. But *Model Cars* can show you the straightaways and hairpin turns of the course ahead.

2

3

*Cars of all shapes and sizes are available to the modeler (1). Among them are detailed kits of the 1934 Rolls-Royce Phantom II in 1:16 scale from Revell (2), HO scale slot cars from Tyco (3) and Jo-Han's 1:25 model of the 1966 Rambler Marlin.*

4

# Getting Started

1

THE TWO HOBBIES in this book—collecting static models for display and buying motorized models for racing—are as different as the types of cars owned by customizers and stock-car racers. Customizers can spend countless hours applying coats of paint to their unique creations; stock-car racers don't care what their cars look like, just as long as they win the race.

If you choose to build and decorate car kits, there are two types to choose from. There are plastic models that are held together with glue, and metal models that are bolted together. The plastic ones can be bought for $2 or less; the most sophisticated metal ones can cost $300 or more. There are two types of models available to the racer also. Slot cars are simple little machines powered by electric motors that run on plastic tracks. Complete sets including track, cars and controls are relatively inexpensive. Radio-control cars are powered by the same kind of internal-combustion engines that are used in model airplanes, and battery-powered electric motors. These cars and the radio transmitters

*The classic car series from Monogram includes a 1:24 scale kit of the Rolls-Royce Phaeton (1). The interior and engine are realistically detailed. AMT's 1:16 scale plastic kit series of Chevy classics produce finished models 12 inches long. One of them is the '57, with vinyl engine hoses and detailed wiring (2). Amazingly detailed kits of racing cars are available from MRC-Tamiya in 1:12 scale. An example of the MRC line is the French Matra MS-11 F-1 (3). Kits in the series feature frames, suspension systems, wiring and cockpit details taken directly from the car's blueprints.*

2

3

used to steer them can cost hundreds of dollars.

Model car building and racing involve thousands, perhaps millions, of people. They include youngsters who snap together the simplest kinds of plastic kits, and skilled and devoted racers who take their cars to the international finals in Milan, Italy.

To help the beginner and to accommodate the experienced modeler, there are various national organizations and associations that provide many services. The International Association of Automotive Modelers and the International Plastic Modelers Society are large organizations that serve the static modeler. Radio Operated Auto Racing, Inc., is a group for the model race-car driver. Each of these groups has plenty to offer to modelers at all levels of expertise.

Another way to find what types of products are available is to visit a large hobby shop. The hundreds of plastic model kits and the wide selection of racers carried by the biggest shops are just a sample of the vast assortment of products sold around the world. In short, there are models to suit

1

*One model in Revell's Chevy Classics series of 1:25 plastic models is the '56, with doors, hood and trunk that open to reveal fine detailing (1). Another classic in 1:25 scale is the '60 DeSoto Adventurer plastic kit from Jo-Han (2). AMT's Trophy Series of 1:25 scale plastic kits includes the '51 Chevy (3). More than twice the size of 1:25 scale kits are 1:12 scale models made by Entex. One of them, the BMW CSL race car (4), features doors, hood and trunk that open, rubber tires, and decals for realistic decoration.*

2

3

anyone's talents and interests, and many tools and supplies to aid the hobbyist. In many cases, the personnel at the hobby shops have used the products themselves and can provide good advice.

All of the kits and race models come with instructions for assembly, tips on their use and often some ideas on ways to display them. For additional advice, there are numerous books and magazines filled with helpful hints from people who have reached a high level of skill.

Model car clubs are other good sources for builders of static models and for racers. They can be found in many cities. Whether there are clubs in your area or not, there will be some serious, experienced modelers or racers who probably will be eager to talk with you and perhaps share some of their special interests in and knowledge of the subject. The owner or operator of a local hobby shop often can direct you to one of these helpful sources. In the case of clubs, you can also find out about many of them and their locations by contacting one of the national organizations that coordinate activities in these hobbies.

As in any serious hobby, planning is very important. Determine first what you want to do: to build from kits, scratch build with the help of some kit parts, collect ready-built models, build and customize racing models, compete in organized racing or any combination of the above. Then you must be sure you have all the necessities: a space to work, the proper tools and the ability to handle the basic costs.

Whichever aspect of the model car hobby you choose, certain basics of information are an important prerequisite. In building, for example, you should know the basic methods of working with specific materials. If you are collecting for fun and profit, you will need to be aware of the elements that make a collectible unique or valuable. In model racing, you will have to have an understanding of the workings of your car's engine, the techniques of piloting it and the rules of the race, to name but a few.

One area of understanding is necessary to all forms of the hobby: scale. It is something you should know about before that very first trip into a hobby shop, and it is something you should fully understand before buying a static model kit or a racing car.

## Scale

Scale is the word used to describe the size of a model in relation to the life-size car. It is expressed as a ratio, such as 1:24, or as a fraction like 1/24th. Either way, it means that one inch of the model

4

represents 24 inches of the actual life-size car. A car that is 192 inches (16 feet) long, for example, modeled on a scale of 1:24 would result in a replica eight inches long ($192 \div 24 = 8$).

One of the most common scales used by manufacturers of static models is 1:25. In addition, models are available in scales of 1:24, 1:32, 1:48, 1:87 and others. Radio-controlled racing cars are normally produced in either 1:8 or 1:12 scale. In 1:12 scale, a 16-foot car will be reproduced as a 16-inch model.

The scales of 1:48 and 1:87 are wise choices if

1

you plan to display your static model cars in a realistic scene. These scales are standard for many types of model trains. Therefore, miniature plastic trees, people and buildings can be purchased in these scales at most hobby and toy stores.

The 1:48 scale is often called 0 scale, and the 1:87 scale is often referred to as HO scale. However, it is important to check the scale ratio on the model package and not go by the 0 or HO designation alone. The reason is that HO means one thing in Europe and something else in the United States. In Europe, HO is 1:76 scale. Also, manufacturers of slot cars often label their models as HO scale, but their true scale is usually 1:64 or 1:66. If you want to build all of your models on the same scale and be sure that all of the miniature buildings and trees you may display with them are in that scale also, be sure to find the correct scale ratio on the box.

Many modelers build all of their cars on a single scale, but others assemble a wide variety of sizes from the 1:8 giants to the tiniest miniatures.

Some of the scale model buildings and landscaping materials used by model car builders are found in the model railroad departments of hobby shops. For that reason, and to help you keep the different scales straight generally, we have put together a chart of the common model car scales and their matching railroad scales. Using this chart, you can, for instance, buy S scale railroad materials for use with your 1:64 scale model cars.

| Model Car Scale | No. Inches To One Foot | Matching Model Railroad Scale |
|---|---|---|
| 1:160 | .075 | N |
| 1:120 | 1/10 | TT |
| 1:96 | 1/8 | E |
| 1:87 | .138 | HO |
| 1:76 | .157 | OO |
| 1:64* | 3/16 | S |
| 1:48 | 1/4 | O |
| 1:32 | 3/8 | No. 1 |
| 1:25 | .480 | — |
| 1:24 | 1/2 | 1/2" scale |
| 1:20 | 3/5 | — |
| 1:16 | 3/4 | 3/4" scale |
| 1:12 | 1.0 | 1" scale |
| 1:8 | 1.6 | — |
| 1:4 | 3.0 | — |

*In the model car world, 1:64 (and sometimes 1:66) is called HO or AFX scale, whereas 1:87 is the common HO scale for model railroads.

No matter what scale you choose, it is important that you do not overestimate your abilities. The best idea is to start with simple kits and gradually work your way up to the more elaborate ones. In assembly of static models and in racing slot cars and radio-control models, many of the skills you acquire will come to you through trial and error. The more sophisticated and expensive the model, the more costly are your mistakes.

2

*Many of the 1:12 scale plastic kits from MRC-Tamiya are suitable for radio-control racing. One of them is the Tyrrell P34 kit (1). Jo-Han makes a plastic kit of the 1964 Cadillac de Ville in 1:25 scale (2). Much easier to assemble is the L.A. Street Machine from Revell (3), one of the kits in the company's snap-together line. No glue or paint is needed to put one of these together.*

3

# Plastic Models

1

2

*Great detail is possible in plastic models, as shown by MRC-Tamiya's 1:12 scale kit of a Tyrrell P34 race car (1). Skill is required to transform those bits of plastic into beautiful models such as those built by craftsmen at Testor Corp. (2). Kits of vintage cars are available from many firms including Monogram, which makes 1:24 scale models of (3, from left) the Rolls-Royce, Lincoln Continental and Packard Speedster.*

3

MODEL CAR BUILDING began shortly after the first automobiles appeared in the early 1900s. The first automobile models were made from scratch by the hobbyist: there were no kits, no pre-formed pieces and no printed instructions, so the miniature cars were fashioned from raw materials. Later, around 1930, wooden model kits became available. These kits usually were nothing more than a set of plans and a box of sticks. It was not until the 1940s that the first plastic kits began to appear on merchants' shelves.

Today, there are all sorts of models to choose from. Classic and antique cars, sports cars, dragsters and stock American and imported models can be found in hobby stores everywhere.

Advances in the formulation of plastics and the manufacture of molds have increased the realism of plastic models over the years. Now the smallest parts can be made with great detail. Modelers often begin by following the kit manufacturer's instructions to the letter, and this can result in a very attractive model. But over a period of time as they become more adept in assembly and finishing techniques, modelers begin to customize and modify kits to create unique works of art.

The simplest kits available are often too simple for adults or older children; the more complicated kits can be a challenge to even the most experienced modeler. It is wise to select a model that may prove to be a bit too simple. This will allow you to acquire a "feel" for the plastic, to acquaint yourself with step-by-step assembly and make mistakes that will not be costly.

1

2

3

4

5

Developing skill in the assembly and finishing of a model is an important part of this hobby, but to some modelers, the construction and painting are little more than the means to an end. A more exciting aspect of car modeling for them is the acquisition of a replica of a favorite automobile; that is, the full-scale car is more important to them than the construction steps involved in building a miniature version of it. This kind of thinking can be likened to that of baseball-card collectors. The cards themselves are not as important to some collectors as are the players they represent.

Much of the fun in the hobby of car modeling comes before the pieces are spread out on the

6

7

8

*One of the newest models in Monogram's lineup is the 1:8 scale 1978 Corvette (1). It is 23 inches long when assembled. An AMT model in 1:25 scale is the '69 Buick Riviera (2). Smaller and less detailed are the snap-together kits from Lindberg like the 1930 Packard Sport Phaeton (3). In 1:12 scale is the new '65 Mustang GT Fastback kit from Revell (4). Lindberg also makes a Snap Fit kit of the Ford Thunderbird (5) that is easy enough for young children to assemble. Fine detailing is possible in the larger scales like 1:16, as shown by the Entex kit of a Packard Convertible Coupe (6). No gluing or painting is necessary with Revell's 1:32 kit of the '56 Corvette (7). A famous car in 1:25 scale is Jo-Han's Chrysler Turbine (8). Also famous is the '57 Ford T-Bird, another Revell snap-together kit (9).*

9

work table, even before the hobbyist goes shopping for a kit. The planning—the thought put into determining which car is needed to fill a gap in the modeler's display—is enjoyable in itself. The first car you buy as a hobbyist may be that 1957 Chevy Bel Air you remember from your youth, now reproduced in miniature by Revell; that fine Jo-Han replica of the '31 Cadillac Town Sedan your grandfather or father daydreamed of; or the class E-Type Jaguar Coupe that you once saw at an auto show and can buy in 1:8 scale from Monogram. We all have our own special interests, and the model manufacturers accommodate these diverse tastes with a wide variety of models.

You may be a Pontiac fan, and decide to acquire a complete set of models from every year of the car's production. You may want to build a kit of every Ford Thunderbird model available, in all scales. Or you may center in on Chrysler products, building kits from every model manufacturer until you have a representation of Chrysler Corporation history in your den. You may want a display of formula race cars, and buy every kit made by MRC.

Some of the more important steps in assembling a plastic car model are taken before the box is opened. Once you have selected the model you want and have brought it home, you should prepare a work area. A large, level area located away from other family activities is best. You will want to be able to work on the model without interruption, and be able to leave it unattended for hours at a time. It is not good to have to pack up a half-finished

1

2

model every now and then, because glues and paints may not dry properly if the model is moved around. Whatever area you choose, be sure to cover the work surface with oilcloth or newspapers to protect it from the glues and paints you will be using. Most of these substances can permanently damage furniture. Be sure, also, to keep these poisonous materials well out of the reach of small children.

Youngsters assembling their first models very often do not bother to sand rough edges from the plastic parts. Glue is often applied straight from the tube and paint goes on with an inexpensive little brush. But as the modeler's proficiency increases along with the complexity and cost of the models purchased, an assortment of tools and accessories are used to enhance the quality of workmanship. Most experienced modelers would agree that there

3

4

5

6

*Very small but nevertheless detailed is Revell's 1:48 scale kit of the '39 Mercedes 540-K (1). A newer Mercedes, from Entex, is available in 1:12 scale (2). One model in Jo-Han's series of classics is the 1:25 scale 1931 Cadillac Cabriolet (3). MPC makes a detailed model of a Chevrolet Camaro in 1:25 scale (4). Another classic Chevy, the '55 convertible, is available from AMT in 1:16 scale (5). A 1:25 scale model of the famous Porsche Speedster is manufactured by Revell (6).*

1

2

3

4

are some tools and materials that even a beginner should not be without. They include the following items.

**Hobby knives.** Razor knives like those made by X-Acto are excellent to work with. They are used to trim rough edges from the model parts and for a variety of other purposes.

**Files, emery boards, sandpaper.** These items can be very useful in removing burrs from the plastic parts. Jeweler's files are used by many modelers who customize their cars. Very fine sandpaper, #400 for example, can be used to smooth glued joints.

**Tweezers and needle-nose pliers.** These can be used to hold small parts for painting or sanding.

**Toothpicks and cotton-tipped swabs.** Toothpicks are sometimes used to apply small drops of glue. Swabs can be helpful in applying paint and decals.

**Clothes pins, masking tape and rubber bands.** These can be used to hold glued pieces together while they dry.

5

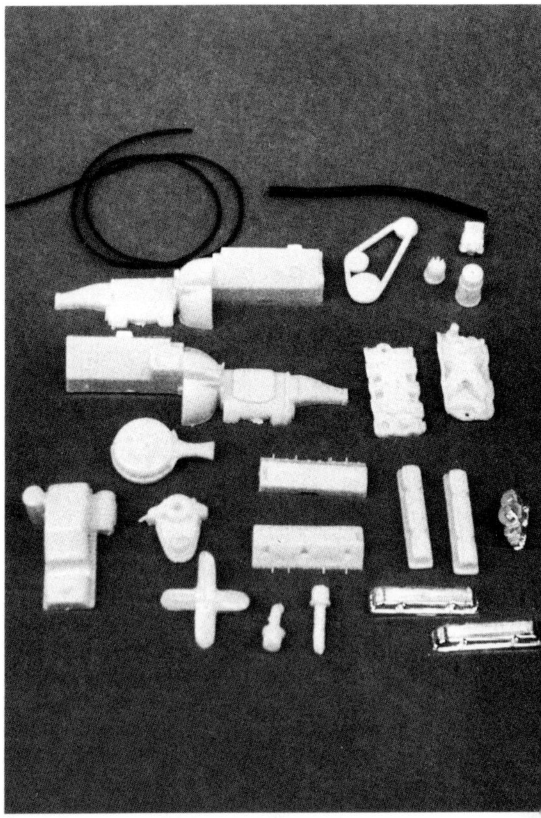

6

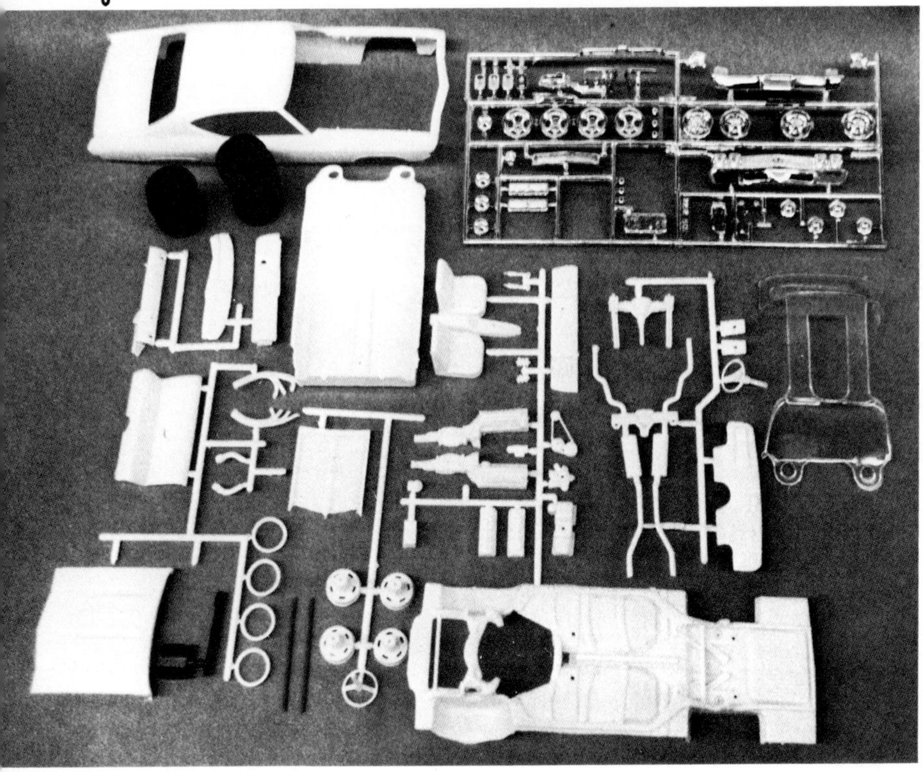

7

*Jo-Han's line of 1:25 scale plastic models includes a Cadillac ambulance complete with detailed stretcher (1). MPC makes a 1:25 kit of the 1978 Dodge Monaco (2). A flawless paint job makes AMT's 1955 Chevrolet an outstanding 1:16 model (3). Also in 1:16 scale is the Entex kit of a 1937 Packard formal sedan (4). AMT's 1:25 scale model of a 1969 Chevy Malibu (5) begins with the plastic parts connected to their "trees" (7). The engine of a 1:16 model is itself made up of many pieces for great detail (6).*

**Scissors.** A pair of scissors is handy for cutting decals and for many other modeling chores.

**Putty.** Special modeling putty is available to fill unwanted gouges in the plastic and to cover seams.

**Paint brushes.** Small brushes used to apply paint details should be of good quality. In choosing them, be sure that they are intended for use with enamels. Water-color brushes used with modeling paints will shed hairs and ruin a paint job.

As you become involved with advanced models and sophisticated customizing, you will undoubtedly find the need for other tools including jeweler's saws, drills and bits, vises and clamps, jeweler's screwdrivers and tin snips. One way of meeting your basic needs for tools is to purchase a chest or kit specially designed for modelers.

### Assembly

At this point, with all the tools and supplies gathered and work area prepared, you are ready to begin the actual assembly. The most important advice to any modeler is: Do not rush.

The main pieces of your model will be separate; smaller pieces normally will be attached to the plastic tree. Check the larger pieces to see if there is any extra plastic on the edges. If there is, trim it away, using a hobby knife, file or emery board. As it becomes necessary to remove pieces from the plastic trees, cut them loose with a knife instead of trying to break them off. Remove only the piece you need at the time you need it. Smooth the edges with sandpaper or a file and then fit the pieces together according to the instruction sheet. Check to see if any adjustment is necessary before applying adhesive.

Styrene cement is generally used with plastic kits. It is available in tubes and in bottles with a small brush attached to the cap. This type of glue actually dissolves the plastic and then bonds it together. Therefore, it can damage any surface of the model that it touches.

Be sure that the surface to be glued is clean and dry; then apply the glue with a toothpick or the

1

2

3

*Painting and detailing are very important to modelers, no matter what kind of model is assembled. Beautiful paint jobs add sparkle to AMT's 1:25 '50 Ford convertible (1) and '56 Fairlane (2), Lindberg's 1:16 1914 Mercedes racer (3), Revell's 1:16 1934 Rolls-Royce (4), the Entex Mercedes 300SL in 1:16 scale (5), MPC's '57 Chevy in 1:25 (6), the Entex '31 Ford (7) and Mercer Raceabout (8) in 1:16, and the '41 Plymouth (9) and '66 Mustang (10) in 1:25 from AMT.*

4

5

brush. Be careful not to use too much. A thin layer generally will bond better than one that is too thick. If the pieces must be held together while the glue dries, use a clothespin, masking tape, rubber band or some other form of clamp. Be sure to allow enough time for the glue to dry thoroughly.

**Painting and Finishing**

You may find that many parts of the model are easier to paint before they are glued to the other pieces. If spray paint is to be used to paint these and the rest of the model, or even if you will use a brush on the smaller pieces, it is helpful to position the pieces on a strip of masking tape. This will keep them from being blown around by the spray.

When the larger parts of the model are assembled and the glue is dry, you can apply modeling putty to any gaps that need to be filled in. When it is dry, sand it smooth with fine-grit sandpaper. You may want to sand the entire model very lightly, to give the first coat of paint a surface that it can grab

6

8

10

7

9

1

2

*Authenticity results from skilled detailing. Models such as the Entex 1:16 scale kit of the 1912 Ford Model T (1) and Revell's 1:48 scale 1930 Packard (2) can be painted to match the originals. Careful use of decals enhance the realism of MRC's Ferrari racer (3).*

3

onto. Whether you sand the model or not, you should wash it with soap and water, rinse it and allow it to dry. From this point on, you should not pick up the model with your bare fingers. Instead, use a cloth or facial tissue to prevent body oil, dust and dirt from collecting on it.

Spray paint goes on evenly and can result in a smooth finish for the car. But it can cause big cleanup problems for modelers who do not take care to prevent the spray from drifting all over the room. For that reason, many hobbyists build a spraying box to contain the spray. Such a box can be made from a cardboard carton with an open top or one open side. The model is placed inside the box for painting. This is an ideal way to reduce the amount of paint that floats around the room and gets into the modeler's lungs. A painting box also helps to prevent dust from settling on the drying model. Be sure you have enough light to see the model well. This will help you to apply an even coat of paint and avoid missing spots.

As you begin to paint, keep the model about eight inches from the paint sprayer. If you hold the nozzle too close, the paint will go on too thickly, drip and spoil the surface. Be sure to move the sprayer across the surface of the model, stopping the spray at the end of each sweep. This will avoid paint buildup at either end of the model. Do all the hard-to-get-at places first: the underside of the body, window and door openings, and other tight spots. You will probably want to apply about five coats of paint to the body to achieve the proper finished look.

### Decals, Details and Display

Paint detailing is best accomplished with very small brushes, such as #1 and #0. These can be purchased at hobby shops or art supply stores. Small bottles of gloss and flat paint in a wide range of colors for accents are part of many modelers' equipment.

To paint lettering on a model car, it is good to make a sketch of the lettering on a piece of paper and use the sketch as a guide to draw an outline of the letters on the model. Then you can paint over the pencil lines.

If you use decals, each one should be cut from the sheet as you are ready to use it. Hold the decal with tweezers as you dip it into some water. Soak it for a few seconds until it begins to separate from its paper backing. Use the tweezers to hold the backing, and a cotton-tipped swab to slide the decal onto the model. A small brush or swab can be used to wipe away excess water and smooth out bubbles between the decal and the model.

The number and type of details you apply to the model are limited only by your imagination.

Many modelers spend as much time in working out ways to display their cars as they do in painting and detailing them. Arranging them neatly on a

4

5

6

7

For many modelers, cementing a model's parts together is a chore--merely the means to an end. With care, the end result can be a beautifully finished work of art, as exemplified by MPC's 1:25 scale Pontiac Firebird (4); AMT's 1:25 scale '58 Chevy Impala (5); three of the cars in Monogram's Early Iron series (6, from left) - the 1958 Thunderbird, 1930 Ford coupe and 1930 Ford Cabriolet; 1:25 scale models of De-Sotos (7, from left), the '60 from Jo-Han and the '56 from X-El; two '59s (8, from left) - the Chrysler from X-El and the Dodge from Jo-Han; and Jo-Han's 1955 Pontiac Star Chief in 1:25 scale (9).

8

9

1

2

3

4

5

6

7

*Corvette fans have a wide range of scales and model years to select from MPC alone: the 1:16 scale 1963 Corvette (1); the 1:32 snap-together '63 Vette (2); and the 1:25 scale kit of the 1960 version, with a hood that swings open (3). Many modelers use plastic display cases like those from Jo-Han (4) to protect their models. Great detail is not available only to modelers who build large kits, as shown by Revell's tiny 1:48 scale kit of the '34 Duesenberg SJ (5). Other Corvettes are made by (6, from left) MPC and AMT in 1:25 scale. Another classic, the '39 Jaguar, is made in 1:16 scale by Entex (7).*

bookshelf is the simplest way, but you may feel that your artistry deserves a more creative showcase. A plastic display case to enclose each model and protect it from dust can be purchased at most hobby shops. They are usually quite inexpensive. To fully display your workmanship, you may want to cover the bottom and back of the case with mirrors to reflect the otherwise hidden details. A shadow-box shelf can be built to display and protect models, with a separate compartment to hold each one in position.

A method of display that has become especially popular in recent years is the construction of dioramas which place the model in a realistic setting devised by the model builder.

To create a simple diorama, you can just mount the car on a piece of plywood that has been painted or covered with something resembling pavement or ground. The car can be roped off with a small nylon cord strung through posts made of aluminum tubing. Thin chain, like that used for a delicate necklace, can be used instead of cord.

If you prefer an action scene, the plywood base can be painted to look like a race track or a highway, and a background scene can be painted and attached to the back of the base. Some pictures from magazines can make good background settings for a model. The pictures can be glued to a sheet of cardboard. Other items such as miniature tires and tools can be added.

When you have collected a number of automobile models, you can make a large display in which all the cars are arranged on a base. One, for example, could be set up to resemble an auto show; manufacturer's booths can be created and added to provide a realistic touch. If the models are racing cars, you can make a race track complete with spectators in a grandstand. An interesting highway scene can be created using plaster hills and cliffs with rocks and moss glued on, trees at the side of the road, and houses and billboards in the background.

# Metal Models

1

*A very elaborate metal kit is the 1935 Mercedes-Benz Cabriolet (1) from Pocher. Simpler and less expensive Monogram kits are (2, from left) a '52 Corvette, '56 Thunderbird and the MG-TC. Martoys of Italy makes a 1:24 scale model of the Audi 80GT (3), and Solido makes a 1:43 scale model of the Ferrari GT (4). The latter two cars are not kits, but many hobbyists repaint such pre-built models. Solido makes metal model kits also.*

2

3

4

IT'S NOT SURPRISING that metal models came before plastic ones. After all, metalworking was perfected centuries ago, but plastics did not come into wide industrial use until after 1909. Early forms of plastic were difficult to work with. As more versatile types were developed, plastics almost completely replaced metal in the manufacture of model cars. Yet metal model kits and metal toy cars are still available, and still are popular with modelers and collectors.

Experience with plastic model car building can be valuable to a beginning metal car modeler, but it is by no means a prerequisite. Some techniques carry over from the one type of modeling to the other, and some metal kits have plastic parts. But basic construction techniques for each type of kit, the tools required and the cost of the models are diverse. A hobbyist who builds metal models often uses a screwdriver instead of a tube of glue, and the price of advanced kits can run into hundreds of dollars. Also, the number of parts in a complex metal kit is far greater than the number of pieces in a plastic one.

Considering that they can be more expensive than plastic kits and more difficult to assemble, and often require special tools, what are the advantages of building metal models? A metal car body can be painted and polished just like the body of a real car. Metal models are heavier and more durable than plastic models. Furthermore, a metal model will hold its value or even increase in value over the years more than a plastic one will.

One reason for the higher cost of metal models is the expense of their manufacture. Pressure die-casting is the most common method of forming the metal parts for these kits.

The first and perhaps most important step in the process of die-casting is the preparation of an accurate drawing of the model after carefully studying and measuring the full-size car to be copied. If the actual car cannot be measured, it is necessary to obtain measurements from other sources like photographs or blueprints.

Once a satisfactory drawing has been made, a pattern must be carved, usually out of wood. (Sometimes the pattern is made from synthetics or other pliable materials.) The pattern is made to a scale larger than that of the desired model. This is done so errors can be more easily detected.

The pattern is then cut in half. Each half is coated with liquid plastic. What is formed is a reverse pattern. At this stage, a special machine reduces the scale and produces a mold to the desired size. The mold is cut from vanadium steel.

The next step is the construction of the plumbing parts: some will transport hot metal to the mold, some will carry water for cooling, and others will carry hot air under high pressure. All plumbing parts are attached to the mold. The mold is then set in a pressure die-casting machine where molten metal is driven into the mold under pressure. The water circulating around the mold cools the metal. The mold then separates and the casting is ejected.

**Availability and Assembly**

Metal model kits are not as easy to find as are plastic kits. Hobby shops carry some, and others are available by mail. Several magazines for hobbyists frequently run advertisements from manufacturers of these models.

Among the simplest and least expensive are the 1:24 scale Metal Master kits produced by Monogram which have die-cast metal body parts and accessory pieces made of plastic.

Gabriel, maker of Hubley metal models, is prob-

5

6

*Burago of Italy makes metal toys and will soon begin production of a series of kits in 1:24 scale, including the Porsche 911-S (5,6).*

1

2

3

*Truly astonishing detail separates the Pocher metal models from the rest. One of them, the 1934 Rolls-Royce Torpedo Cabriolet Phantom II is a beautiful example of the modeler's art from the spoked wheels (1) to the miniature instrument panel gauges and foam padding in the seats (2) to the famous Flying Lady hood ornament (3). It all goes together with nuts and bolts and a lot of skill.*

4

ably the most widely known name in metal car modeling. The company offers a variety of metal car kits, mostly in a scale of 1:20. The models vary in complexity from those having approximately 35 parts to models that contain more than 120: a Gabriel model of the classic Duesenberg has over 34 parts for the engine alone.

The most elaborate, however, are the models from Pocher of Turin, Italy. Pocher's kit of a Rolls-Royce Cabriolet Phantom II, a 1:8 scale replica of a car built for the Maharaja of Rajkot in 1934, contains more than 2000 parts. When assembled, it makes a model more than two feet long. Pocher offers a wide variety of very expensive models of the great classic cars.

Whether you assemble the simple kits from Monogram, the extremely advanced kits from Pocher or any metal kit in between, you will find it necessary to use several tools that builders of plastic models do not need. Files capable of smoothing rough edges of metal pieces are required, as are screwdrivers of several sizes. Other tools that are helpful include small wrenches, Allen or hex wrenches, needle-nose pliers, a jeweler's saw, an oil can with a pinpoint nozzle, tweezers, a small vise and a soldering iron.

In the metal kit you buy, you will find detailed instructions for assembly. Study the instructions and familiarize yourself with all of the pieces. Be careful when removing parts from the kit, because it is very easy to lose small pieces. Try to work in an area where you will not be interrupted or have to pick up the parts each time you stop assembly. Be sure to keep all the small pieces safely in one place: a misplaced screw or nut can cause a frustrating delay.

Custom painting is an important aspect of the metal car modeling technique. Most metal model kits have parts that are unpainted; therefore, fine finishing is left solely to the modeler. Preassembled models usually come already painted, but many modelers often are not satisfied with the quality of these paint jobs and choose to paint over the original factory finish. A model may be carefully put together and flawlessly detailed, but it can be a disaster if the painting is inadequate. Fortunately, mistakes of this type can usually be corrected; but a lot of headaches and heartaches can be avoided if the paint is applied properly the first time around.

Metal models, unlike those made of plastic, can take automotive lacquer and modeling lacquer as well as enamels designed specifically for models. The best results are generally achieved by using paints that are made for the hobby of modeling. They are thinner than others and tend to go on more evenly. Therefore, they are able to highlight details instead of obscuring them as ordinary paints can and often do.

Prepare a metal model for painting just as you would a plastic model. Clean the model or parts to be painted thoroughly to remove all dirt, dust, oil and grease. A good way to prepare a metal surface for painting is with a good application of a metal cleaner and conditioner, like that made by Floquil. Once the piece is clean, be sure not to touch it with anything other than tweezers or a lint-free cloth. Oil from your fingers will prevent the paint from adhering properly to the metal.

The techniques of painting metal models and parts include both brushing and spraying. Careful brushing can be a most effective form of painting models because it often produces an extremely smooth, uniform coating. This depends, however, on the skill and care of the painter and on the quality of the brush. Inexpensive brushes can be sloppy and inadequate for modeling purposes. To achieve good results, you may be able to use camel's hair or red sable top-quality brushes, like those made by Grumbacher. Be certain to check with the hobby shop or art supply store to determine whether the brush you select can be used with the paint you want.

For spray painting, which can also be an effective method if it is done properly, there is a choice of equipment: disposable spray cans, airbrushes and spray guns. Good-quality airbrushes are best, but they can be quite expensive; a modeler's airbrush with accompanying air compressor, like those from Badger or Bammco, can cost more than $100. There are, however, relatively inexpensive ones on the market which will do a satisfactory job in the hands of a careful modeler.

All parts that are being painted the same color should have each coat applied at the same painting session if practical, no matter which technique —brushing or spraying—is used. This will aid in the eventual uniformity that you are seeking and it will enable you to inspect and compare (and perhaps correct mistakes) as you go. Preassembled models may have to be repainted without being taken apart, but if they can be disassembled, you will be able to achieve a better overall paint job.

In working with metals, the best method is to first apply a primer coat and than add several thin coats of paint. After the primer coat and between each subsequent coat of finishing paint, the model or parts should be allowed to dry thoroughly. When you have the color, luster and overall uniformity you want, you are finished with the paint job.

There may be other details you want to add, including the decals and decorations often used by plastic-model builders. However, some metal materials made for general hobby use can be adapted to customize a metal car. Such things as metal tubing and sheets can be shaped to fit.

Metal models are handsome show pieces. They can be strikingly displayed alone, in groups or in a diorama. No matter how you display your models, you probably will want to provide some kind of protection for it. Transparent plastic cases are inexpensive. Glass cases are available ready-made or can be specially built to fit your models.

# Collectibles

1

*Beginners in the hobby of collecting model cars sometimes can be fooled by modern copies of antique models. In the case of Julian Thomas, a modeler who recasts and reissues cast-iron cars, it is easy to tell which is which. Thomas' cars, including three early Studebakers (1), are stamped with his name and a notice that they are copies of the originals. Another type of collectible is the promo model, such as those made by AMT in the 1:25 scale (2). Such promos line the walls of collector Ronn Pittman's display room (3).*

2

TRY TO IMAGINE positioning enough model cars bumper-to-bumper to form a parade one mile long. If each car were a foot in length, you would need 5280 of them; but few models are that big. Our estimate of the number of cars it would take is 15,000. The same number of full-size cars would form a procession more than 35 miles long.

Now try to imagine all of those model cars being owned by one person—in many cases, assembled and painted by that one person—and displayed in his house. That is what you could see if you visited the home of Ronn Pittman, car collector extraordinaire. You would see models in the living room, in the family room, in the basement, the garage and everywhere else. You would not see all 15,000 because some are carefully packed away for safekeeping. But you would see enough model cars in Pittman's midwestern home to get a rough idea of the time, effort and expense that can be devoted to the hobby of model car collecting. Pittman's one-man traffic jam boggles the mind.

But Pittman, as most serious model car collectors, does not buy cars with the goal of setting records. Quantity is not as important to Pittman as quality. Each item satisfies a particular desire on the part of the collector and is added to fill a

3

specific gap. Collecting can be an art, a highly individualized matter given direction by each collector's tastes and resources.

Some modelers collect one type of model: cast metal, for example. Others devote their efforts to collecting on a theme: the entire line of Ford cars from the Model A to the Fairmont, perhaps. Still others may want only antique or classic cars, or foreign-made cars, or racing cars.

Most model car builders are collectors, regardless whether they ever buy a commercially made miniature or trade the results of their labors for some other sought-after model. The only model car builders who cannot be called collectors are those who give away or otherwise get rid of each model car after it has been built. Needless to say, such modelers are few in number.

However, collecting often goes beyond merely saving every completed kit, progressing beyond the realm of a simple hobby. It can be a solid investment, and it will be if you go about it in the right way. It can also be costly, especially if your ambitions are on the same level as Pittman's. A collection of 15,000 metal and plastic models—many antique, many unique—could strain the budgets of most hobbyists.

1

2

3

*Toys often make up a large part of a model car collector's display. Many such toys, including die-cast models of early Graham automobiles from Tootsie-Toy are extremely rare and valuable (1). Not all model cars collected are made of metal or plastic. Some, like the Hudson Miniatures Model T, made in the late 1940s, are made of wood and cardboard (2). Tootsie-Toy's die-cast models of a 1935 Ford with trailer and 1935 Chrysler Air Flow are not greatly detailed, but they are valuable nevertheless (3). Other cast-iron models reissued by Julian Thomas roll on white-rubber wheels, just as the originals did (4).*

4

When does a group of models painstakingly put together, painted and detailed become a true collection? Where would you begin if you were to decide to start a collection? If you have preserved the models you have built, you already are a collector. To start at the very beginning, you simply go out and buy a few models and build them with care; or you purchase some ready-made collector's items. This procedure can be simple or complex, depending on how much of yourself and your money you want to put into it.

You will need to know what kinds of cars are collectible; that is, the types of models that are valued by other collectors. There are three general areas of miniature cars that collectors are interested in. The categories overlap somewhat.

### Kinds of Collectibles

**Model cars built by a hobbyist:** This includes all kit and scratch-built models (or a combination of the two), from the earliest days of modeling in the 1920s to present-day models. This group includes only scale models or exact replicas of full-size vehicles. The materials they are made of include cast iron, cast aluminum, wood, plastic, or a combination of these materials. One example would be the elegant scale-model kits produced today by Pocher.

**Pre-built models:** These are the commercially built models (sometimes called "factory-built" even though they may not have been constructed in a factory) that an individual may buy with all details, paint and other decorations already fashioned. The category includes scale models and exact replicas as well as facsimilies that were made for any purpose other than as toys. These miniatures are made of cast metals, wood, plastic, glass and ceramics. Examples are the Corgi or Matchbox models on the market today, and the "promos"; that is, promotional models displayed or given away in the past by automobile dealers or created to promote other products.

**Toy cars:** This category covers the entire range of antique toy cars that were crafted of quality materials—cast-iron bodies and rubber wheels, for example—in the years prior to World War II. (Some toy cars built in the late 1940s and the 1950s are also collected but they are not nearly as sought-after by collectors.) Antique toy cars often command the highest prices in today's collecting market, sometimes ranging into the hundreds of dollars for a unique specimen in good condition. They are not collected only by car buffs and hobbyists but also by people who are interested in antiques of all kinds. Examples of these collectibles are the famous Tootsietoys, early Dinky models, Hudson Miniatures and Arcade Toys. Some are powered by friction motors; others have free-rolling wheels.

5

6

*Ronn Pittman's collection includes promos, metal models and toys. One promo is a model of a VW Microbus made in Germany (5). The metal BMW model is from Burago (6), and the toy '39 Buick is from Dinky (7).*

7

1

2

3

The first step in seriously approaching this hobby is to determine what to collect: not merely what you want to collect but what is worthy of collection in terms of the item itself and its relative value. Old or antique items of any kind guarantee value because they are generally rare, out of production and therefore irreplaceable; and they contribute to our enduring sense of nostalgia. This is very true of model and toy cars.

When does a model or toy rate the classification antique? The answer to that varies from collector to collector. World War II traditionally has been the time period that model car collectors have used to separate old models from contemporary ones. The year 1940, in fact, is often accepted as the precise line of demarcation. The reason for this separation into two eras is that the war served as an interval between the time when almost all models and toy

4

5

6

7

8

*A '49 Kaiser in 1:20 scale (1) was made in the 1950s. A '49 Hudson in 1:16 scale was produced in limited quantities for dealer use (2). Plastic promos made by Cruver, such as the '49 Oldsmobiles (3), are valuable 1:25 models because so few were made. Other promos sought by collectors include the '59 Rambler Metropolitan from Hubley (4), and the '57 Ford Custom 300 (5), '54 Buick Skylark (6) and '58 Pontiac Star Chief (7), all from AMT. The '59 Rambler Ambassador Wagon and '60 Rambler Classic (8) are from Jo-Han.*

cars were made of cast metals and the time when plastic emerged as a dominant material of manufacture. Antique hunters and collectors often disagree with one another on this distinction, with many holding that only those toys and models produced before 1920 qualify as true antiques.

The following is a generalized breakdown of the history of model and toy collectibles that you can use as a practical guide for your particular purposes as you explore this hobby.

1900 to 1910: The first scale model autos and replicalike toy cars are made in Europe, especially in France and Germany.

1910 to 1920: The first cast-metal models begin to appear in the United States, mostly produced by automobile manufacturers for design and promotional purposes.

1920 to 1940: Cast-metal toy cars and pre-built

*Also collected are car banks, like the die-cast 1950 Chevrolet coupe made by Banthrico in 1:25 scale (1). Another promo in 1:25 scale is the 1951 Henry J toy from Lincoln Line (2). It has a recoil motor that is wound by pushing the car backward. When released, the car moves forward. Plastic promos in 1:20 scale are the '50 and '51 Plymouths made by Product Miniatures (3). Also in 1:20 scale is the wind-up 1951 Packard Henny hearse from AMT (4). The fact that the delicate swan hood ornament is still in place after almost three decades is an indication that the model has been cared for. According to collectors, the '53 Willys 1:25 scale promo from Burd Manufacturing (5) is extremely rare.*

1

2

3

4

5

models flourish in the United States. Kits are introduced and begin to enjoy growing popularity near the end of the 1930s.

1940 to 1945: World War II; much production ceases, but some cars are manufactured from rubber and wood. Interest in building models and even in collecting all but disappears.

1946 to present: The hobby of building models re-emerges. Plastics begin to replace other materials in the late 1940s and completely dominate the industry by the early 1960s. Collecting grows in popularity in proportion to the renewed interest in model building.

To determine the collectibility of a toy car or model, you have to know something about it. There is no hard-and-fast rule or specific guidebook to use in determining whether a particular model has value in terms of investment potential or rarity. You will have to take these things into consideration and come up with the answers for each item you find on the market or decide to seek out.

To help, we have devised some very basic and important questions that should be asked about any potential collectible, and we have offered some answers and guidelines that should provide a basis for assessing the quality and/or value of a toy or model car.

### Seven Questions for Collectors

*Was (or is) the miniature car hand-built, or was it a product of a factory or other commercial pre-builder?*

This determines the category that the item falls into. It is also a measure of its uniqueness. One-of-a-kind models are obviously the ultimate in rarity, but not necessarily in value. There are some one-of-a-kind models that are nothing more than crude playthings of interest only to their makers. The quality of handcrafting is, of course, a major factor and includes such considerations as authenticity in scale, attention to and rendering of details, painting and other decoration, and materials used.

*If the item is a pre-built model, was it mass produced or was it distributed in limited quantities?*

This will have a distinct bearing on rarity. Many modern-day models, like those from Matchbox, for example, are produced by the thousands. As a result, their value is limited. Promos, too, must be looked at in this light. Some were released to dealers as showpieces and were produced in very restricted quantities. Others, like the mass-produced, plastic "auto-banks" put out by Chevrolet in the late 1940s and early 1950s (the ones that bore the motto: "to help you save for a rainy day . . . or help you buy your new Chevrolet") were produced for distribution to thousands of potential Chevy customers.

*Is the model of pre-World War II vintage, or is it a more recent product?*

6

7

*Many collectible promos also bear the Jo-Han name. One of them is the 1958 Cadillac Fleetwood (6). The 1961 Pontiac Bonneville (7), 1962 Pontiac Tempest (8) and the 1963 Ford Thunderbird (9) are from AMT. The T-Bird, a sport roadster, is an extremely rare and valuable promotional model, much sought after by collectors.*

8

9

1

*Of the 1963 Buick Riviera (1), the 1964 Pontiac GTO (2) and the 1964 Mercury Breezeway (3) from AMT, the rarest is the Mercury. Such plastic promotional models are valued by collectors, who often swap and buy models to obtain a particular promo replica that will complete their displays. Collectors say the value of a promo reflects the value of the original full-scale car. For instance, promo models of Nash cars are rare, but Nash cars were not popular and therefore their promos are not as highly valued as other promos that are not as rare.*

2

3

This basic guide to whether the item is antique or not can be a decisive factor in determining the present and potential worth of the car. Oldness in itself does not constitute value, but it seems that almost all nonperishable items become collectibles if they are kept around long enough and remain in relatively good condition. Beware, however, of inflated prices just because an item is old or is judged to be an antique by a dealer or shop owner. Do not, on the other hand, disregard models and toys being produced today. If they are of high-quality materials and craftsmanship and/or are unique, they will be tomorrow's antiques. Pre-built models such as those from Burago, Politoys and Brooklin or elaborate kits such as those from Pocher and Gabriel (formerly Hubley) are sought after today by knowledgeable collectors.

*Is the collectible in mint condition or are there flaws that detract from its visual impact?*

As in any form of preserving and collecting, only those items that are in reasonably good condition are worth keeping (excluding, of course, those items of historical significance). With almost all old models and toy cars, there will be flaws, missing parts, and other damage resulting from use and handling over the years. If you are buying from a true collector, however, the piece may have been kept in superb condition or have been professionally restored to that state.

When you judge an item as a collectible, be able to determine whether you can restore it yourself. Parts will not be readily available for older pieces unless you can switch them from one model to another. Repainting an old model or toy is generally frowned upon by collectors and will certainly detract from the value of any model. Like an old book that is most valuable in its original dust jacket, an old model should wear its original paint. But certain repairs and refinishing are within the range of almost all modelers and can be the means of turning an inexpensive piece of junk found in a thrift shop into a prized piece for collection. This area of consideration also should be a guideline to the collector as to the importance of protecting and preserving all the models in a collection.

*Can you determine the authenticity of a model or toy car?*

This is no longer an easy task. Since prices have risen in recent years, especially in the area of true antiques, a number of fakes or counterfeit models have made an appearance in the collector's world.

This does not include the legitimate reproductions of old or antique miniatures that are being made today by many reputable model builders,

4

5

6

*Toys like the model of a '50 Cadillac (4) are valuable. Many collectors buy kits like AMT's three-in-one kits from the '60s (5) and save them. AMT produced a model frame for Buick in 1960 (6). Jo-Han's Chrysler turbine cars (7) were seen at the Seattle World's Fair. Ceramic models promote Jim Beam liquor (8).*

7

8

# Ultimate Collectible: Diamonds, Rubies, Silver and Gold

using materials and techniques identical to those used in the originals. These usually are engraved with a notice that they are reproductions. Because of the careful craftsmanship and quality materials, these new models are valuable in themselves as collectibles.

There are, however, counterfeits to watch out for. One guide to authenticity is the name or identification of the maker, usually engraved or inscribed on the original model. Counterfeiters often do not reproduce these markings well—sometimes not at all—but they can be forged.

1

2

*As much a tribute to the artistry of the jeweler, silversmith and goldsmith as to model making is the replica of a 1903 Model A Ford created by master modelers of the American Silversmiths Guild in 1978 to commemorate Ford Motor Company's 75th anniversary.*

*The model, 13 inches long, seven inches wide and eight inches tall, weighs 7.5 pounds. About 85 ounces of that is sterling silver—even the screw fittings are made of silver—and a half ounce is pure gold. Two blue-white, full-cut diamonds are set in the headlamps, and a full-cut ruby is used for the rear lantern.*

*The model was made from 400 separate pieces. It is not an exact scale model, but what it lacks in authenticity it makes up in intricacy. The brake pedal connects to the rear wheels and will actually stop the model when it is moving; the steering wheel turns the wheels; the springs function; the glove compartments open and close; the door at the back of the Model A's tonneau (rear compartment) opens; the shelf below the front seat can be opened to provide a view of the forward portion of the car's engine, gas tank and water tank; and the starting crank can be inserted and turned just as cranks were used to start up the original Model A.*

*The elegant adornments of the model, from the pure gold lettering of the Ford name in its authentic and famous script to the precious gems commissioned from world-renowned jeweler Cartier, make this possibly the single most expensive model car to be found today.*

*Only 1708 units were made to coincide with the number of original Model A cars manufactured by Ford. Each model is numbered and registered and was offered to Ford dealers at a price of $2975.*

*The American Silversmiths Guild has produced numerous commemorative models in precious metals over the years. To illustrate the investment potential of its models, the Guild notes what happened to the selling price of the Fisher Body Coach which was issued in an edition of 1000 units in 1973. The coach sold for $2500 when new. Today, the Guild says, each one commands a price of more than four times that amount.*

White rubber tires once were an immediate identification of an antique model (later replaced by black rubber, then plastics and other synthetics), but they also have been counterfeited lately. Always check the riveting of an old model. If the body has screws or bolts as we know them today, beware. Axles of old were solid nail rivets, not the hollow pins found on many modern fakes. Old models for the most part were characterized by the use of round-head rivets instead of slotted screws.

One of the best ways to protect yourself from being cheated is to deal with reputable collectors,

3

4

*Some collectors work to obtain a complete series of one model line, such as the Chevrolet Corvette. One collection of this type includes an MPC kit model of the '56; a Monogram metal kit of the '53 (1); AMT promos of the '58 and '62 Vettes (2); and a promo of the '63 (3) and one of the '67 (4), also from AMT.*

dealers and stores who will stand behind what they are selling or trading. The other protection is to know the subject well: study it and discuss it with experienced collectors and increase your own expertise.

*Who is the maker of the toy or model?*

This is an important and elementary consideration for the collector. Many names have been established over the years as creators and manufacturers of quality toy and model cars. Their products are the items to look for. There are other excellent and valuable collectibles that carry no name or brand, but they are the exceptions. Here is a list of companies and trade names, both American and foreign, whose model cars and toy cars are sought by collectors. Most of them have ceased operation; but a few, including Gabriel, Corgi and MPC, are still in business. This list is by no means complete, but the names here are the most prominent in the field and should serve as an effective reference source for all collectors.

American La France
Arcade
Barclay
Brooklin
Buddy "L"
Burago
C.R.
Carlisle & Finch
Corgi
Dent
Distler
Dowst
Erie
Fador
Gabriel
Hafner
Hubley
JEP
Jordan
Kelmet
Keystone
Kingsbury
Lehmann
Manoil
Martoy
Matchbox
Mercury
MPC
Politoys
Republic
Rio
Schieble
Strauss
Sturditoy
Tootsietoy
Wells
Wilkins
Ziss

AMT
Auto-Pilen
Bing
Bub
Bugatti
Burnett
Carette
Converse
Dayton Friction
Dinky
Doll
Dugu
F.V.
Freidag
Gunthermann
Hessmobil
Hudson Miniatures
Jo-Han
Kansas Toy & Novelty
Kenton
Kilgore
Knapp
M & L
Märklin
Marx
Mebetoys
Minic
Pocher
Pressman
Revel
Savoye
Solido
Structo
Tipp
Turner
Wiking
Williams

1

2

3

*Where do you go to find collectibles?*

This is no problem if you are collecting contemporary pieces, because they are available from hobby shops, toy stores, mail-order dealers and others. But old or antique models often go unnoticed in dusty attics and dingy basements. This, then, is a question that even the most experienced collectors ask one another.

4

5

*Promos are made for foreign firms also, like Volvo (1,2) and Saab (3). Collectors' items are made of plastic like the '59 and '60 AMT Edsel promos (4); or pewter, like the Cadillac Seville (5). Some collectors want a promo for each year, like AMT's Corvairs from '62, '60 and '61 (6).*

6

We went to master collector Ronn Pittman for an answer. "Scour the house sales, garage sales and auctions of antiques and household goods," he said. "Go to the rummage sales, thrift shops, antique stores, junk shops, flea markets. Check out the automobile dealers and talk with them: if they aren't involved with promos now, they may know sources who are. Talk with hobby shop owners. Seek out any model car clubs near you. Check the ads in modeling magazines. And make it known that you are serious about collecting."

Model car collecting requires dedication, and a commitment of time, learning and money. But it is one of the most satisfying aspects of the model car hobby. In the long run, it can prove to be a lucrative investment as well.

# Slot Car Racing

1

*One of the Matchbox Speedtrack sets from Lesney features 12-1/2 feet of track and cars in HO scale (1). Another HO slot racer, from Tyco, is chrome plated (2). Cox makes slot car sets in a scale larger than HO. One of them, called the Big 8, comes complete with two "formula" cars and enough track for a four-foot layout (3).*

2

3

GREAT INTEREST in slot car racing and many commercial race tracks sprang up around the United States in the 1960s. Inside a racing shop you could see and hear crowds of slot car fans grouped around each of several miniature roadways, cheering the competitors and waiting for their turns at the controls. The walls of such buildings often were lined with shelves full of car kits, tools, paints and tires; and enthusiasts stood at the counters, carefully inspecting various components in hopes of improving the performance of their cars. Slot car racing is no longer the fad it once was, but the cars and organized competition are still available.

Some slot car racers take their hobby very seriously, spending a great deal of time and effort in building unique cars from scratch and fine-tuning them for each hotly contested race. For them, there are numerous kits as well as a full supply of separate parts that can be used to fashion a prize-winning car. Other slot car fanciers, such as young children and their parents, buy complete sets which include assembled cars and pieces of track. There are several brands of sets available for these rec room racers.

Sets sold by Aurora, Tyco and other slot car manufacturers are made in HO scale. This is one of the popular scales for home slot car sets. (Remember, when talking about slot cars, HO scale is 1:64 or 1:66 – not the model train HO scale of 1:87.)

HO racing, or AFX, as it is called by Aurora, is often most practical for home setups because HO systems are small and the tracks can be set up in a variety of space-saving configurations. Long, relatively narrow rectangles and seemingly endless patterns of loops that thread their way under and over one another can provide a lot of racing distance in a minimum of square footage.

Another fact important to many home slot car racers is that HO cars are much cheaper than the larger, more sophisticated models that come in 1:32 and 1:24 scales (the ones generally raced on commercial tracks). The HO systems (track, rheostats, wiring, etc.) are much less expensive than larger systems and are much easier to set up and maintain.

Still another desirable feature of HO is that there is such an incredibly wide variety of racing cars to choose from, at various prices. In nearly all cases, one brand of car can be used on a different brand of track. Beyond that, many manufacturers (like Aurora) produce racing cars so their chassis and bodies are interchangeable as well.

If you begin with a complete HO set, as most HO racers do, you will have all the basics in one large box. As your interest and racing expertise grow, however, you may soon move beyond what a basic set can offer. You will want faster cars, perhaps modified and customized by your own hand. You can take existing model slot cars and add wider tires for better traction or weights to give it better handling. Or you may want to install a better set of

4

*Rheostats control the speed of Ideal's HO cars (4). Tyco makes a "grandstand" that produces racing sounds (5).*

5

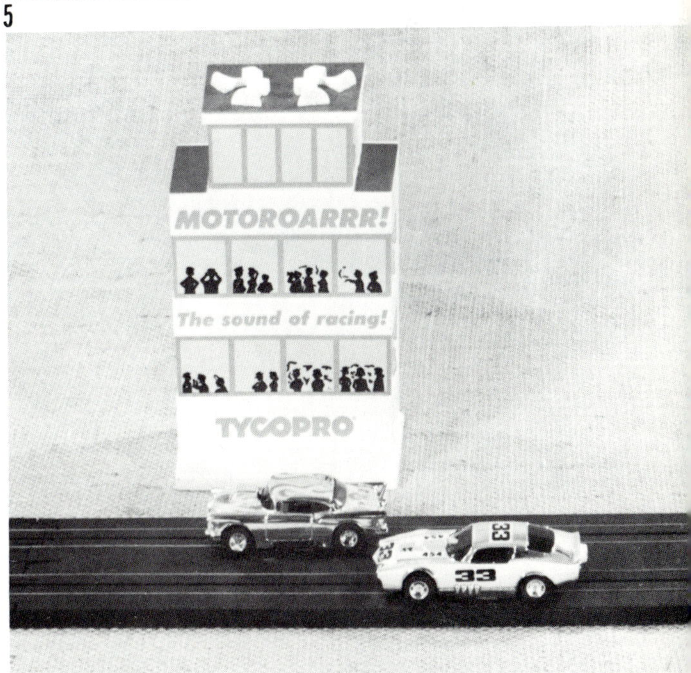

1

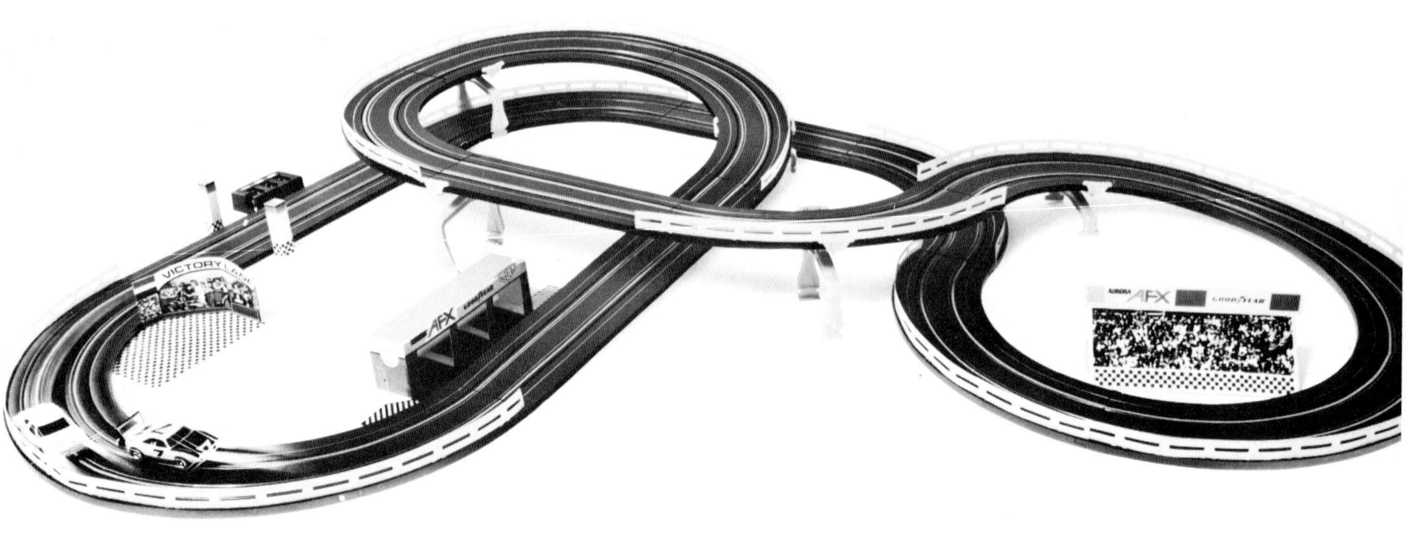


2

gears and motor brushes.

Slot cars are painted and decorated at the factory. Many hobbyists, however, prefer to repaint their cars and add their own markings and identifications. This process of customizing is a simple one. There are many model paints (enamels and lacquers) on the market that will provide realistic colors and lusters if they are applied correctly. The same principles of painting plastic and metal model cars described elsewhere in this book can serve as your guide to effective repainting of slot cars. You can paint right over the original paint or, if you prefer, you can strip the model down to its base and start fresh. Lettering, numbering and various other decals in scale are available at most hobby stores.

Track systems can be enlarged and redesigned in

3

4

*The essential elements of any slot car set are the power pack, track, car and hand-held speed controls such as those made by Tyco (1). TCR sets from Ideal feature a lane-changing switch and a jam car that the other racers must pass successfully to keep racing (2). One of the larger slot car sets from Matchbox is the LeMans 300, which includes 20 feet of track and cars with lighted headlamps (3). HO slot cars such as those from Tyco are available in many different colors (4). Slot cars and tracks in HO scale are popular with many hobbyists who do not have enough area available in their rec rooms to set up larger-scale courses.*

an almost limitless number of configurations. You can recreate the Watkins Glen Grand Prix Raceway or the Daytona Raceway, for example, as well as other well-known race courses. There are guides for this in books you can find in libraries or bookstores, and in the *AFX Road Racing Handbook*, published by Aurora.

One of the best ways to maintain your HO slot car setup, once it is as complete as you want it, is to make it permanent. The track can be secured to a table top, large wall shelf with supporting legs or simply to a large piece of plywood. The environment around it can be made realistic with landscaping; and model buildings, pit stops, billboards, grandstands, gas pumps and figures. Many of these are available from slot car manufacturers; others

1

can be adapted from what is available in the model railroader's world of scenic pieces, all of which can be found in hobby shops and toy stores. You may also build your own environmental features from scratch. There are many tips and guides for this in the books and magazines devoted to model car racing.

To keep the cars running well, it is important to properly maintain them and the track. A covering for the entire track system can be fashioned from plastic, paper, old sheets or blankets. It should be used to protect the setup and keep it clean when not in use. Track should be cleaned often and carefully. There are track cleaners available at most hobby stores. The racing cars must be kept clean and well lubricated. Instructions for maintenance often are included with slot cars, whether they are purchased separately or in kits.

Most of the slot cars that come in kits and the ones built from scratch are generally in the larger

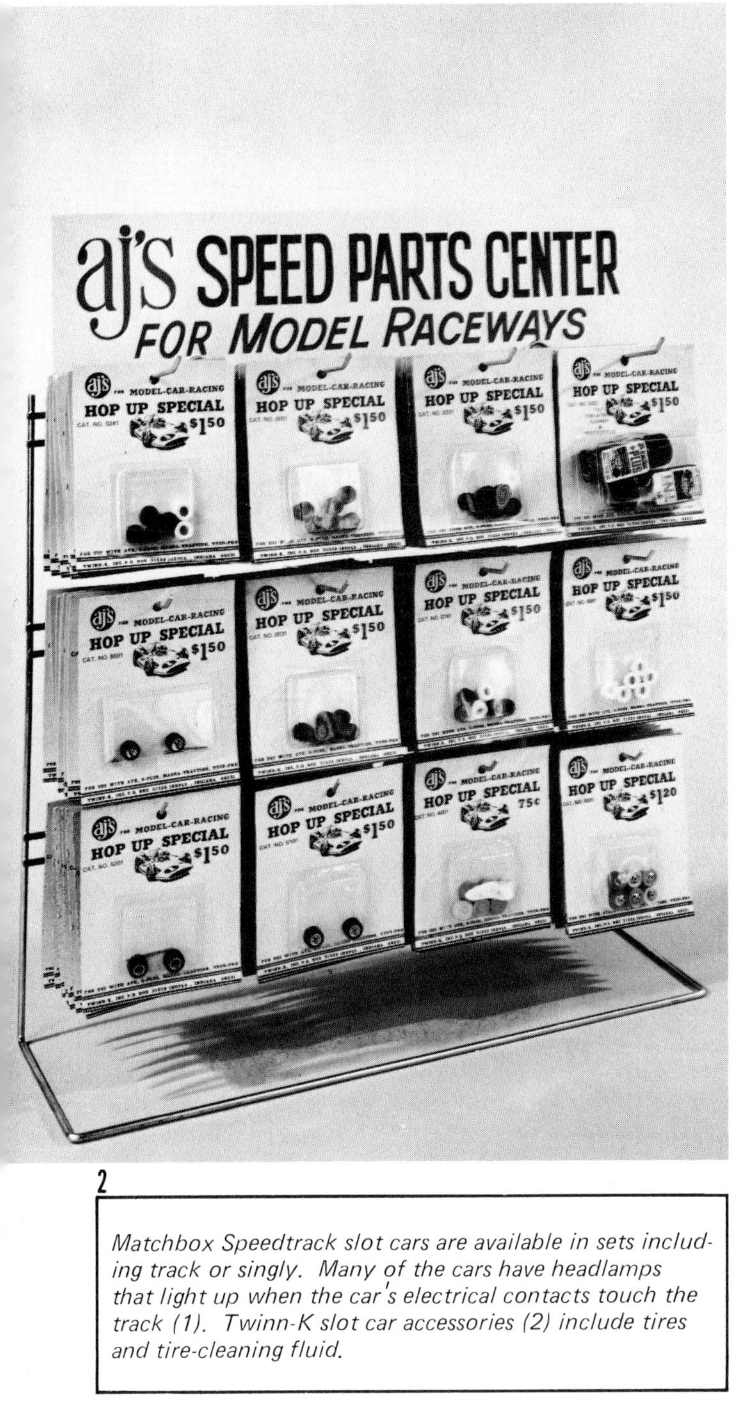


2

*Matchbox Speedtrack slot cars are available in sets including track or singly. Many of the cars have headlamps that light up when the car's electrical contacts touch the track (1). Twinn-K slot car accessories (2) include tires and tire-cleaning fluid.*

scales of 1:32 or 1:24. These are also the most popular scales for racing at slot car centers and in organized competitions (both scales can be raced on the same track). The 1:24 scale today is thought to be fastest and is the choice of many experts, but the 1:32 scale is considered the international racing scale and is preferred by some clubs in the United States for organized racing.

Slot cars of all scales operate in the same way. They all have small electric motors. They run on direct current, so a transformer and a rectifier are needed to change household or alternating current to DC. Complete sets normally have this transformer built into them.

The tracks the cars run on, which are usually made of plastic, have a slot in each lane. This slot is about 1/8 inch wide and 1/4 inch deep. A pin or fin called a guide flag on the underside of the car fits into the slot to help hold the car on the track and to keep the car's electrical contacts in position to pick up electrical current. The car's contacts are made of braided metal wire or a similar material. As the car is set on the track, its contacts touch metal strips that run along the track on either side of the slot. Current flows through wires from the transformer to the track's metal strips, to the contacts on the car, and from there to the car's electric motor. A system of gears drives the rear wheels of the car.

There are some types of electrically powered model cars that run on tracks without slots. These cars can change lanes and be steered to some degree. The cars we are talking about here, true slot cars, cannot be steered; however, their performance on the track can be controlled by the operator using a hand-held device. This device, called a rheostat, regulates the amount of electricity that flows through the metal strips in each lane. By using this control, the operator can speed up the car on straightaways and slow it down as it approaches curves. Some hand controls also have brakes; that is, switches to block the circuit altogether and stop the car quickly.

**Building Cars and Tracks**

If you purchased a slot car set with assembled cars and raced them for awhile, you may have found that your racing abilities soon progressed beyond the capabilities of the cars. If so, you may want to build a high-performance model from a kit or from scratch.

Most of the kits available contain everything you will need to build a slot car except paint and tools. They also have detailed instructions which provide step-by-step guidance as to how the pieces go together. Among the pieces included will be a steel and brass chassis, a motor, steel axles, tires, gears, electrical contacts, a clear plastic body and decals.

Soldering is a very important part of building a good slot car. The type of solder most often used for wires and parts has a tin/lead ratio of 60/40 and has a resin core. Silver solder with a low melting point is also popular because it provides additional strength. Regardless of the type of solder you use, you will need a good soldering gun or iron that is heated to the proper temperature. It is also important that the metals to be joined be completely clean. Sandpaper or a file can be used to smooth and clean the metals. After sanding, be sure to wipe the parts thoroughly. Also be sure the tip of the soldering iron is clean. In slot car building, not

1

*Big Bank is the name of the top-of-the-line slot racing set from Cox (1). The cars, bigger than HO scale, can both race in the same lane. Tyco's HO scale cars (2) are smaller than those from Cox, but are just as exciting to race.*

2

much solder is needed to join pieces of metal or wiring, but to be effective the solder and the two pieces being soldered must be heated equally.

The solder should cool slowly. Do not blow on it or apply drops of water to hot solder, because this can cause it to crack. As a last measure of success, be sure to scrub the joint with toothpaste or wash it with diluted ammonia if you have used acid-core solder or flux.

The Lexan plastic bodies made for slot cars are very light and practically indestructible. Paints specially made for Lexan plastics can be purchased at most hobby shops.

One thing to inspect periodically is the car's wheel alignment. Make sure the chassis is aligned and the axles are not bent. Also make sure the tires are round. Check to be sure that electrical contacts are clean and working properly. If the contacts are too stiff or become loose, they may cause your car to pop out of its slot. The motor of the slot car should also be kept clean so that it can operate at maximum efficiency.

These are just the basics for the slot car builder. There are many innovations in construction, design adaptation and engine refinement that the creative hobbyist can incorporate into his or her slot car.

Such refinements are necessary if the car is to be raced in organized events. The Hobby Industry of America's Model Car Racing Division can provide you with a list of these competitions. The clubs and other organizations that sponsor these and other races have developed a set of specific guidelines which must be followed by all entrants. Among other things, the rules state: all slot cars entered must not be more than 3-1/4 inches wide; all four wheels must touch the track and roll; only one guide flag can be used on a car; and a three-dimensional figure of a driver must be located behind the steering wheel.

Another phase of the hobby and an additional challenge is to build your own race track. A half-inch-thick surface of Masonite or particle board is most often used for the track because it is smooth and strong. You will have to cut your own slot.

After the slot is cut, the track should be painted with a latex paint. This must be done before the contact strips are applied. When the paint is thoroughly dry, copper or aluminum tape or braid is attached to each side of the slot with contact cement or epoxy.

After that, you can add decorations. Hobby shops usually carry a variety of miniature trees, buildings and various other realistic items of scenery. To make hills and mountains, you can use window screening for a base and cover it with papier mâché.

The last step is to hook up the wires from the transformer. This is not difficult because there are only a positive and a negative wire to attach to each lane on the track. However, it is important to solder all the wires to avoid loose connections.

# Radio Control

*The Martini Porsche 935, a detailed model from MRC, is a four-cell electric racer for use with RC equipment.*

IN THE PIT AREAS at trackside under a bright spring sun, drivers and mechanics are fueling up their cars and checking all the last-minute details. One by one, the engines are started; amidst the din and clouds of exhaust the sleek Indy-type cars begin to move out onto the oval track. The drivers inch their automobiles into position on the starting line and watch intently as a race official touches the tip of his green flag to the pavement. Suddenly, he raises it, and the cars roar off toward the first turn with each driver straining and challenging the others for the lead.

One car is stuck at the starting line with engine problems. The driver jerks and punches the controls in vain: the engine sputters and dies. Disgusted, the driver reaches down to the car, scoops it up in his hands and walks away.

Meanwhile, the race continues. Each driver

1

twists and turns the little steering wheel on his control box and manipulates the throttle lever to accelerate and back off on the gas, and these operations are instantly transmitted to the car by radio waves. This is radio-control, or RC, racing. It is the closest you can get to actual auto racing without climbing in behind the wheel of a life-size automobile. The cars are only about two feet long and less than a foot wide, but the realism of their behavior on the track can generate the same kind of competitiveness and excitement as the racing of full-scale cars.

RC model car racing enables you to put your artistic, mechanical and driving skills to the test; yet it leaves you free of the threat of injury and saves you the thousands of dollars required to buy and maintain a life-size racing machine. Radio-control racing, which has been in existence as a sport for only about 10 years, is a fast-growing hobby in the United States and abroad.

It offers the greatest amount of manueverability and control of any type of model car racing because the car is not encumbered by slots, wires or tracks that go with other forms of model car racers. An RC car can be operated on almost any hard or paved surface and tracks can be set up in a variety of places. RC cars powered by internal-combustion engines usually must be raced outdoors, but electric-powered cars often are raced indoors as well as outside. RC model car building and racing is a more expensive hobby than other forms of model car building and racing, but it allows a lot more direct participation by the hobbyist.

RC car racing also has an advantage over another popular pastime, model aviation. After the initial expense for the car, engine and electronic apparatus, the cost of operating an RC car in terms of fuel consumption is minimal compared to that of model airplane flying.

The cars, like model airplanes, can be bought ready-made, with only slight assembly needed before racing. Like airplanes, they can be purchased in kit form, and some are so simple that they can be put together in less than a day. Some experienced

3

4

The Peerless Pop Buggy (1) is a gas-powered RC model. MRC's 1:12 scale Porsche racer can be displayed with detailed figures (2). Pro-Cision makes a battery RC racer in 1:24 scale called the Lancia Stratos (3). A 1:8 Peerless racer is the Super Dash (4).

2

RC car modelers, like some model aviators, build their racers from scratch; however, the beginning hobbyist would be wise to gain some experience with a ready-to-run model or a simple kit before progressing to the much more difficult project of building such a car.

Despite the relative ease of assembling a ready-to-run or simple kit car, these racers do not suffer any handicaps in competition with those built from scratch, because they are designed by engineers with years of experience in radio-control racing. So RC events are truly open to those who are less interested in building a car than in winning races.

### Scale and Power

Radio-control cars are most often built on a scale of either 1:8 or 1:12. The 1:12 scale, the smaller of the two, has a proportion of one inch to one foot; that is, a 1:12 model of a 15-foot car would be 15 inches long. This scale is readily available in both preassembled and kit forms. A 1:12 scale car is usually about 14 inches long, six inches wide and weighs about 2-1/2 pounds. It is also the scale most often recommended to beginning hobbyists, since it is less expensive and less complicated than 1:8 scale models. The cost of a 1:12 scale model car with an appropriate engine and radio-control system is generally about half as much as that of a similarly outfitted 1:8 scale model. Cars in 1:12 scale can attain speeds of 35 miles per hour, not as fast as 1:8 scale models which have been clocked at better than 50 miles per hour, but certainly fast enough to offer an exciting challenge to the driver.

The 1:8 scale model cars, the largest and fastest of RC racers, usually weigh between five and six pounds and are approximately two feet long and 10 inches wide. This scale is the one most favored by model car hobbyists who decide to build their models from scratch. There are many parts for these cars available separately: different bodies, brake systems, clutches, and other components.

There are some RC cars on the market today that

1

2

Bolink's line of RC racers in 1:12 scale includes the Datsun 280-Z (1). A clear plastic body and precision-machined metal parts make up the Peerless Super Dash (2). Galoob makes several RC racers, including the DMW and the Corvette Stingray (3), controlled by a very small hand set. Ideal's line of battery-powered racers with fully proportional steering includes the DK Pro-Am Special (4).

3

4

5

*Other 1:10 scale RC racers in the Peerless line are the Peanut Wagon Buggy (5) and the Peanut Dune Buggy (6). They use gas engines and two-channel radio systems.*

6

are built on scales between 1:12 and 1:24. They are smaller than 1:8 scale and operate differently. These cars, usually much cheaper in terms of cost and construction than larger ones, have impulse-type radios rather than digitally proportional radios. This means they cannot be controlled in terms of speed or their degree of turn. They can only respond to their maximum; that is, full right or full left. This type of car is generally considered a toy, not a hobby model like the larger-scale RC racers and not appropriate for racing with the larger ones.

The internal-combustion engine used in model planes is the kind most often used in RC cars, adapted to drive a flywheel and centrifugal clutch instead of a propeller. This type of engine burns a fuel which is similar to kerosene. Batteries and an auxiliary motor are needed to start it.

The size of these power plants is measured in cubic inches of displacement, the volume of the cylinder in which the piston moves up and down. Power output is directly related to displacement, so one engine with greater displacement than another could be expected to be more powerful. An RC car on the 1:8 scale usually needs an engine with a displacement of 0.19 to 0.21 inches, while a smaller 1:12 scale racer can use a 0.049 or 0.050 cubic-inch engine.

The 1:12 scale cars also can be powered by electric motors that use quick-charge batteries. These motors are slightly more expensive than model airplane engines. They require about 15 minutes for a complete charge of their batteries. They offer the advantages of simpler operation and the ability to be used indoors. A radio-control car with an electric motor does not require an auxiliary starter motor or messy fuel. You simply flip a switch and it is ready to go. Electric-powered and gasoline-powered RC cars are equally fast. Most racing clubs hold events for both kinds.

The two types of electric motors that are classified for RC racing are the four-cell and the six-cell sizes. Four-cell motors, using four batteries, can reach speeds up to 22 miles an hour and are designed mainly for indoor racing. A six-cell car can be used either outdoors or indoors and is usually the selection of the more serious or more experienced hobbyist. It can attain speeds up to 30 miles per hour. (A four-cell car motor can be converted to a six-cell, but it will need a more powerful charger and the proper batteries.)

**Radio-Control Systems**

Remote-control radio equipment for model cars is available in a broad range of styles and capabilities. Some radio-control systems are made specifically for model cars; others are built for model airplanes but can be adapted for use with cars. This radio equipment is purchased separately, and can cost approximately as much as the car and its engine. Once purchased, though, a remote-control box can be used with more than one car.

An RC system basically consists of a transmitter, a receiver and servomotors. The transmitter is the control panel that the driver holds in his hands to operate the car. It usually transmits radio signals on two or three channels, which means it can control two or three different functions. Using a two-channel unit, the driver can control steering on one channel and acceleration on the other by manipulating controls on the hand-held panel.

This works in much the same way as a regular radio does when it plays music. Just as a broadcasting station sends out a music signal to your radio that causes the mechanical parts in the speaker to move, the RC transmitter sends out a signal from its antenna to the one on the car and causes the servomotors to move. These little motors, in turn, activate the mechanisms in the car that control its steering, accelerator and brakes.

Just as your home radio must be turned to the proper frequency of the station you want to listen to, the RC transmitter and car must be tuned to the same frequency. The frequency used by a particular car and control panel is determined by the matching crystals in the transmitter and receiver. If you wish to operate more than one car at a time, you must use a different frequency and different crystal for each car. If you wish to change frequencies, you must change the crystals. This is usually a simple plug-in job. Crystals can be purchased at most hobby shops and through the mail.

There are 19 different frequencies presently available for RC model car racing. Generally, unless a specific frequency is requested, the dealer who sells you the RC system will select its frequency for you.

Unlike RC model plane flying, RC car racing normally does not require a license from the Federal Communications Commission. The usual power of an RC transmitter is under 100 milliwatts, which is sufficient to control a model car as long as it remains in sight of the operator. These units do not need licenses. However, a stronger radio which can provide better response does require licensing by the FCC. An experienced RC hobbyist or the personnel at a hobby shop can help you determine how powerful a radio system you need. If you decide to buy a unit that operates on a level above 100 milliwatts, an FCC form may be included in the equipment package. You can also obtain the required form, FCC form 505, directly from the FCC.

### Racing

There are many RC racing clubs and groups across the country that sponsor races and can offer guidance in operating and racing model cars.

Radio Operated Auto Racing, Inc., is the national organization that has set up specific rules and regulations for 1:12 and 1:8 scale model racing. In addition, ROAR sponsors races all over the country and holds national championships as well.

Since there are 19 frequencies available to model car racers, 19 cars can be raced at the same time; however, this is seldom the case because the size of the track will usually limit the race to fewer competitors.

Radio-control equipment and cars have increased in sophistication over the years. The cars are now more responsive—for example, most now have digital proportional steering instead of the less exact type of steering we mentioned earlier—and therefore require more skill on the part of the race driver. These advances make racing a lot of fun, but they also mean that the driver must often practice maneuvering his or her car to train for competition and remain alert during a race.

RC racing requires a bit more than just bringing your car to a track and racing it. Because these cars are fast-moving projectiles, there are a number of safety requirements to be met and regulations to be followed. These guidelines include: positioning boards around the track to prevent the cars from spinning off and injuring someone, keeping spectators away from the track, and assuring that the race is covered by sufficient insurance (the latter is a ROAR requirement).

Officials are another necessity and may include all of the following: a race director, a registrar, a technical inspector, a frequency controller, a starter, timers/lap judges and timers/lap counters. ROAR can provide you with full safety guidelines for its sponsored events.

Races are held in three categories of skill: novice, amateur and expert. This opens the sport up to almost everyone. RC racing is available to the hobbyist on local, regional, national and now international levels. Sanctioned races are held in more than 50 countries. Throughout North America, the sport is growing at an ever-increasing rate. Today's RC racing involves all kinds of model cars: Indy-type, CanAm, GT, stock cars, drag racers, fuel dragsters and funny cars. The types of races include oval track, road races and drag races.

Here are some general guidelines for conducting the various types of RC competitions. From these guidelines, you can get an idea of what is involved in organized, competitive RC racing. By following them, you probably can set up your own racing meet or event. If you are truly serious about it, however, you ought to obtain the latest issue of ROAR Racing Rules.

**Oval Track Races:** The track should have two straightaways (comparatively long and a minimum of 20 feet apart), connected by two arcs. The track surface should be uniform and preferably made of coarse concrete or unsealed asphalt. Barriers should be placed at all turns and in any area directly adjacent to spectator areas. Boundaries should be clearly marked with stripes. Drivers' areas should be elevated and afford equally clear view of the track area for all contestants. Oval races are normally run in a counterclockwise direction.

**Road Races:** There is much more leeway here to develop your own race course, which will depend on the size of the racing area and its location. Any road race course, however, should have a series of left and right turns and at least one long straightaway. Road surfaces should be the same as for oval racing, and boundaries or course markings should be clearly visible. All drivers should have equally good views of the race course and adequate space in which to function (a minimum of four square

1

There's much more to the LaTrax Cobra RC racer (1) than meets the eye. An independent floating axle in the rear (2) provides road-hugging traction. The gearing and suspension systems used in front (3) allow the car to lean into turns. The operations of the racer are controlled by a complex receiving module under the body (4). LaTrax RC cars use battery power. Pro-Cision's 1:16 scale Maserati Bora (5) also is electrically powered. Radio-control racers from Cox, however, are powered by internal-combustion engines. One Cox model is the Interceptor (6).

4

2

3

5

6

feet). All spectator areas and drivers' posts should be protected with barriers.

**Drag Races:** The drag strip should consist of clearly marked straightaways with minimum lengths of 165 feet for 1:8 scale racing cars and 110 feet for 1:12 scale racers. A stopping area at the end of the drag strip must be at least 100 feet long for 1:8 cars and 75 feet long for 1:12 scale models. Surface material should be concrete or unsealed asphalt. Barriers should protect spectator and driver areas. Races should be divided into events for dragsters and funny cars, and the cars should be generally and immediately recognizable as such to qualify in either category. No tire traction additives should be allowed. Crossing of a lane boundary or jump starts are causes for disqualification in a particular heat.

The rules for entrants in a race; and factors affecting the cars, related equipment and the drivers can be set up on an individual basis. ROAR's specific requirements, restrictions and regulations enable a uniformity of competition. It may not always be possible to strictly follow all of ROAR's rules, but that should not deprive you of the fun and excitement of RC racing.

Competition includes more than racing: there are *Concours d'Elegance* competitions in which cars are judged on scale authenticity, body craftsmanship, exterior finish, working features and other specifics of design and construction.

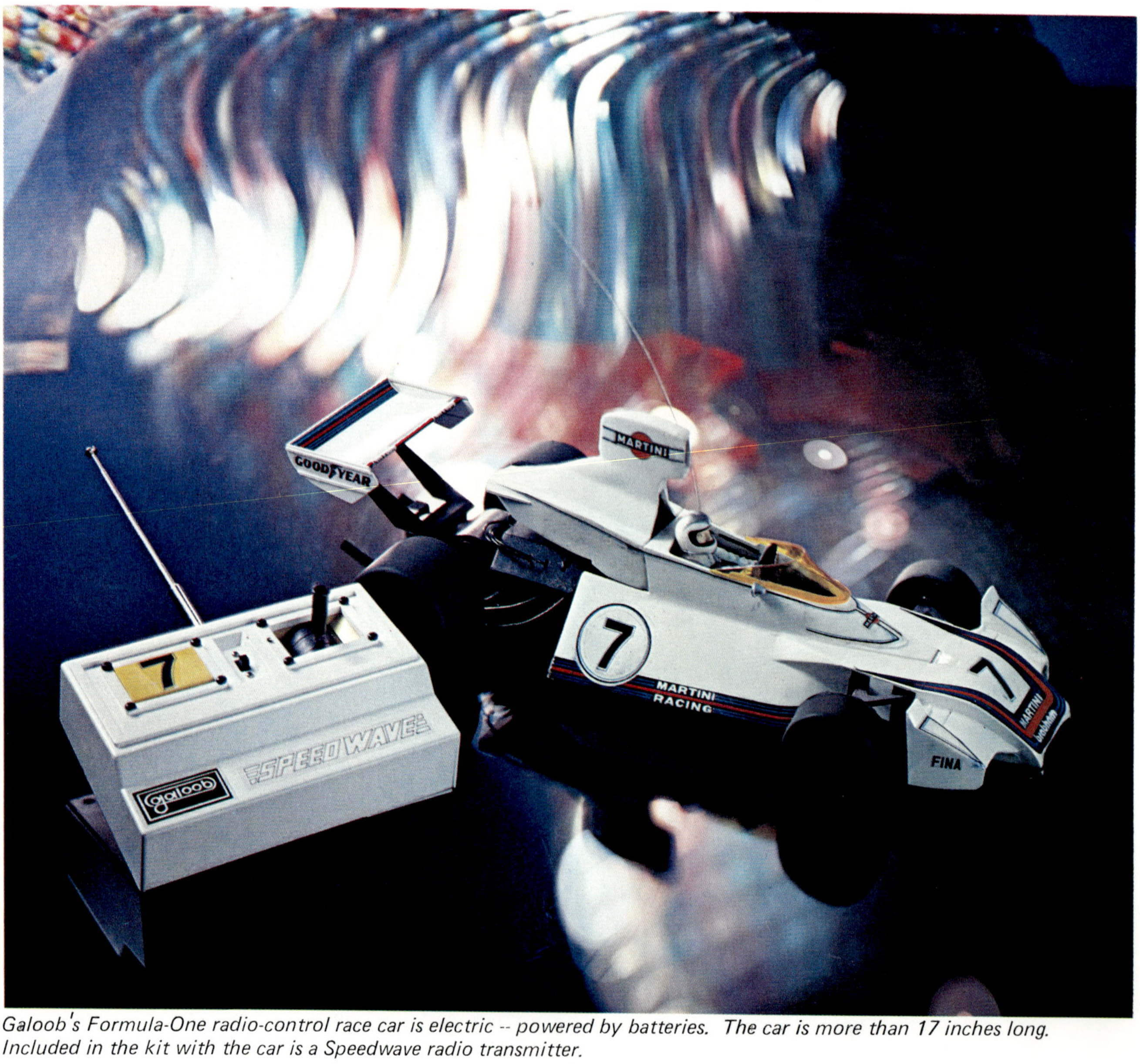


*Galoob's Formula-One radio-control race car is electric -- powered by batteries. The car is more than 17 inches long. Included in the kit with the car is a Speedwave radio transmitter.*

# Sources

## Associations & Organizations

A NUMBER OF national organizations and associations are dedicated to the interests of the model car builder and the model car racer. Each offers various services, and most of them have their own publications to help you keep informed of what is going on in the world of car modeling and racing. They are excellent sources for an exchange of ideas and for coordinating and controlling competitions of many kinds.

**Federal Communications Commission**
1919 M Street, N.W.
Washington, DC 20554

This is the governmental agency that issues licenses for all radio communications and transmissions in the United States. You will need a license from the FCC if the power of your RC apparatus is above 100 milliwatts.

**Hobby Industry of America**
Model Car Racing Division
319 E. 54th St.
Elmwood Park, NJ 07407

Slot car fans will find this organization helpful, because the group can provide a list of centers all around the United States where you can race your cars.

**International Plastic Modelers Society/USA**
P.O. Box 2555
Long Beach, CA 90801

This national organization, with numerous chapters throughout the country, devotes its efforts to all areas of plastic modeling: cars, airplanes, boats, military miniatures and others. Model airplanes seem to get more attention than cars, but IPMS/USA can still be of interest to the model car builder. The organization provides a variety of services to the static modeler: two periodicals, *Update* and the *IPMS/USA Quarterly* are published; and the organization also sponsors competitions, contests and a national convention. About 5000 members currently belong to IPMS/USA. Dues are $10 a year for adults and $6 for those under 18. That fee includes subscriptions to both publications.

**Model Car Collectors Association**
c/o Mark Gustavson
7524 S. Crest Circle
Midvale, UT 84047

The MCCA, more than 1000 members strong at present and with an average monthly increase in membership of from 6 to 8 percent, serves the collector and the builder of model cars. Its heaviest emphasis is on plastic models. This is an excellent organization for the truly interested car modeler to get involved with. Its 15 or so chapters across the country offer good opportunities for modelers to get together and meet others with similar interests. The MCCA sponsors a variety of competitions, setting down rules and regulations, scoring methods and coordinating activities. Another benefit of membership is a subscription to the MCCA's bimonthly magazine, *Model Car Journal.* Membership dues are $10 a year.

**Radio Operated Auto Racing, Inc.**
20860 Homeland Rd.
Matteson, IL 60443

For the RC model car racer, ROAR is practically a must. It is the national organizing and governing body for the sport and it offers a wide range of services. ROAR publishes a bimonthly newsletter *Rev-Up,* which brings members the latest news of races and other events, race results and information about equipment. It also publishes a set of rules for car and race track specifications and racing conduct. The national organization is broken down into eight U.S. regions, each with its own director and coordinator. Within each region are a number of individual clubs that hold their own contests and meetings. Also, a ROAR-sponsored national championship is held each year. Liability insurance for race organizers is also available at low rates to ROAR members. The price of individual membership is $5 a year; family membership is also available with annual dues of $10.

# Publications

THE FOLLOWING are some prominent and helpful publications for the serious car modeler and/or racer. They offer a wide range of guidance and information about current events in the world of model cars.

***ABC's of Model Car Racing,*** Auto World, 701 N. Keyser Ave., Scranton, PA 18508. This good overview of all aspects of model car racing—RC and slot cars—is one of the most reasonably priced introductions to the sport available: 25 cents at the time we went to press.

***International Modeler,*** Sensory Perceptions, P. O. Box 1208, Topanga, CA 90290. A slick, good-looking magazine aimed at the adult modeler, this publication pays a lot of attention to model cars but also covers other modeling fields as well. Well-illustrated articles and a large classified ad section should be of interest to any serious modeler. It is published bimonthly and is available by subscription and at newsstands and hobby shops.

***IPMS/USA Quarterly,*** International Plastic Modelers Society, P.O. Box 2555, Long Beach, CA 90801. This magazine is highly informative. Well-illustrated articles cover the entire range of plastic modeling, with car modeling getting its appropriate share of attention. As its name suggests, the magazine is published four times a year. It comes free with membership in the IPMS/USA.

***Racing Circuits,*** LH Publications Inc., 289 Quincy Place, RD No. 1, Langhorne, PA 19047. "Journal of radio-controlled car racing" is what the publisher calls this magazine/newspaper, and it is truly one of the foremost publications in the field. It includes wide-ranging articles on happenings and trends in the sport of RC car racing, plus good coverage of competitions. It is available by subscription and at some hobby shops.

***R/C Buyers Guide,*** Clifton House, Clifton, VA 22024. R/C Buyers Guide is an excellent source book for the more than 2000 RC products on the market today. It is a publication that a model racer should refer to before investing in RC equipment. It

is well worth the few dollars it costs.

***Rev-Up Newsletter,*** ROAR Inc., 20860 Homeland Rd., Matteson, IL 60443. This bimonthly newsletter is the official publication of ROAR Inc., and is therefore an important item for any serious model racer. Each issue contains hints and suggestions to help the modeler, plus up-to-date information on race schedules, race results and new equipment to keep the racer on top of events in the sport. It comes with membership in ROAR.

***Scale Modeler,*** Challenge Publications, 7950 Deering Ave., Canoga Park, CA 91304. A monthly magazine, *Scale Modeler* is devoted to all forms of plastic static models, with some good, in-depth articles on model cars from time to time and a variety of tips and hints for the serious modeler. It is available by subscription, at newsstands and at hobby shops.

***Update,*** International Plastic Modelers Society, P. O. Box 2555, Long Beach, CA 90801. *Update* is published every other month. It covers all areas of plastic modeling, and provides many tips and helpful suggestions. Product reviews and current news on contests, conventions and other activities make this publication topical and informative. You automatically become a subscriber when you join IPMS/USA.

***Jerobee Newsletter,*** 12702 N.E. 124th St., Kirkland, WA 98033. This newsletter is put out by JoMac Products, Inc., a large manufacturer of RC racing cars and equipment. A subscription comes automatically with membership in JoMac's Jerobee Race Club. It goes beyond the usual manufacturer's or dealer's newsletter, providing a lot of information about racing schedules and results, and new products from JoMac.

***Model Car Journal,*** Model Car Collectors Association, 7524 S. Crest Circle, Midvale, UT 84047. Aimed directly at the model car builder and collector, this is the official publication of MCCA. It is published bimonthly and contains an interesting assortment of articles, tips, columns, new product reviews and classified advertisements. It is available by subscription only, but the subscription also gives you charter membership in MCCA. Serious modelers will get a lot here for a small investment.

***R/C Sportsman,*** P. O. Box 11247, Reno, NV 89520. This magazine/newspaper is aimed at RC modelers who build or race cars, airplanes and boats. It covers a broad range of interesting RC topics as well as related news. Plenty of advertising and new product reviews will keep you abreast of what is new on the market. Published monthly, it is available by subscription and at many hobby shops.

***AFX Road Racing Handbook Vol. 2,*** Aurora Products Corp., 44 Cherry Valley Rd., West Hempstead, NY 11552. As complete a guide as you could want for slot car racing, this 114-page booklet covers everything from fine-tuning cars to building model slot cars yourself. Articles, tips and product information are extensive. Layouts for building replicas of nine major auto racetracks in the U.S. are included. At its modest price, it is a worthwhile investment for the serious slot car hobbyist.

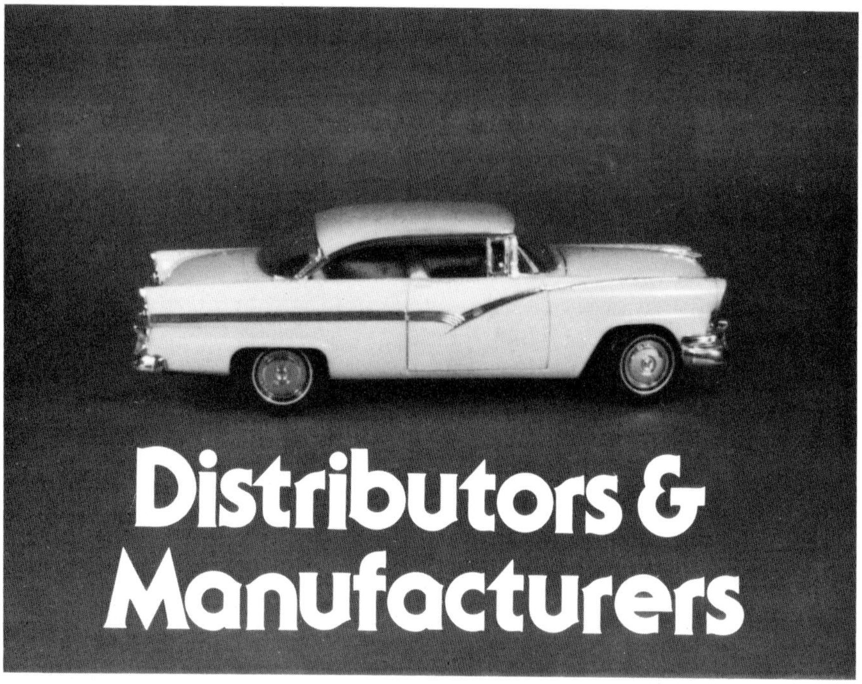

# Distributors & Manufacturers

DISTRIBUTORS AND manufacturers of modeling products provide a number of valuable services to the hobbyist. Most of the companies listed below publish catalogs, many of which are very beautifully illustrated and informative. These can give you an idea of what is available and how much you can expect to pay for it. Many of these firms also publish how-to booklets and other materials that provide helpful hints about the use of their products.

## STATIC MODELS

All are manufacturers or distributors of plastic kits unless otherwise indicated.

**Airfix** (see Ava International)

**AMT**
1225 E. Maple Rd.
Troy, MI 48084

**Associated Hobby Manufacturers**
401 E. Tioga St.
Philadelphia, PA 19134
(Metal and plastic models)

**Aurora Products Corp.**
44 Cherry Valley Rd.
West Hempstead, NY 11552

**Ava International**
P.O. Box 7611
Waco, TX 76710

**Brooklin Models**
Box 339
Brooklin, Ontario
LOB ICO, Canada
(Metal models)

**Brumm** (see Polk's-Brumm)

**Burago** (see Euro Imports)

**Corgi** (see Reeves)

**Dinky** (see Ava International)

**Dugu**
**Sinclairs Auto Miniatures, Inc.**
3831 West 12th St.
Erie, PA 16505
(Metal models)

**Entex Industries Inc.**
1100 W. Walnut St.
Compton, CA 90220

**The Ertl Co.**
805 13th Ave., S.E.
Dyersville, IA 52040

**Euro Imports**
19H Gardner Rd.
Fairfield, NJ 07006
(Metal kits and toys)

**Fundimensions**
26750 23 Mile Rd.
Mt. Clemens, MI 48045

**Gabriel Industries Inc.**
41 Madison Ave.
New York, NY 10010
(Metal models)

**Heller Models** (See Aurora)

**Hubley Metal Classics** (See Gabriel)

**Jo-Han Models Inc.**
17255 Moran Ave.
Detroit, MI 48212

**Life-Like Products, Inc.**
1600 Union Ave.
Baltimore, MD 21211

**Lindberg Products, Inc.**
8050 N. Monticello Ave.
Skokie, IL 60076

**Minicraft/Hasegawa**
1510 W. 228th St.
Torrance, CA 90501

**Model Products Corp.**
126 Groesbeek Hwy.
Mt. Clemens, MI 48483

**Monogram Models Inc.**
8601 Waukegan Rd.
Morton Grove, IL 60053
(Metal and plastic models)

**MPC** (see Fundimensions)

**MRC-Tamiya**
Model Rectifier Corp.
2500 Woodbridge Ave.
Edison, NJ 08817

**Nichimo** (See Minicraft/Hasegawa)

**Otaki** (see Scale Craft)

**Peerless/Kyosho**
3919 M St.
Philadelphia, PA 19124

**Pocher Metal Models**
(see Associated Hobby Manufacturers)

**Politoy**
**Sinclair Auto Miniatures, Inc.**
3831 West 12th St.
Erie, PA 16505
(Metal models)

**Polk's-Brumm**
314 5th Ave.
New York, NY 10001

**Reeves International**
1107 Broadway
New York, NY 10010

**Revell, Inc.**
4223 Glencoe Ave.
Venice, CA 90291

**Scale Craft Models Inc.**
8735 Shirley Ave.
Northridge, CA 91324

**Solido** (see Euro Imports)

**Testor Corp.**
620 Buckbee St.
Rockford, IL 61101

## SLOT CARS

**Aurora Products Corp.**
44 Cherry Valley Rd.
West Hempstead, NY 11552

**Bill Steube Specialties**
1108 Wardlaw St., East
Long Beach, CA 90807

**Carmen Enterprises**
P.O. Box 381
Greenbelt, MD 20770

**Champion of Chamblee**
5620 New Peachtree Rd.
Chamblee, GA 30341

**Cox Hobbies**
1505 E. Warner Ave.
Santa Ana, CA 92702

**Elmsford Raceway**
380 North Saw Mill River Rd.
Elmsford, NY 10523

**Ideal Toy Corp.**
184-10 Jamaica Ave.
Hollis, NY 11423

**Joel Monteque Cars**
P.O. Box 381
Greenbelt, MD 20770

**Lesney Products Corp.**
141 Commercial Ave.
Moonachie, NJ 07074

**Lionel Powerpassers**
26750 23 Mile Rd.
Mt. Clemens, MI 48045

**Matchbox** (see Lesney)

**Mura**
1630 162nd Ave.
San Leandro, CA 94578

**Parma International**
4651 W. 130 St.
Cleveland, OH 44135

**Phase III**
826 Conklin St.
Farmingdale, NY 11735

**Tony Przyblowicz Parts**
8 Crescent Lane
Irvington, NJ 07111

**Twinn-K Inc.**
P.O. Box 31228
Indianapolis, IN 46231

**Tyco Industries**
540 Glen Ave.
Moorestown, NJ 08057

## RACING CARS

**Ace R/C Inc.**
P.O. Box 511
Higginsville, MO 64037

**AMT**
1225 Maple Rd.
Troy, MI 48084

**Aristo-Craft Distinctive Miniatures**
314 Fifth Ave.
New York, NY 10001

**Associated Electrics**
1928 E. Edinger Ave.
Santa Ana, CA 92705

**Autotrol**
Auto World
701 N. Keyser Ave.
Scranton, PA 18508

**BoLink Industries**
P.O. Box 80653
Atlanta, GA 30366

**Burago** (see Euro Imports)

**Cox Hobbies, Inc.**
1505 E. Warner Ave.
Santa Ana, CA 92702

**Curtis Dyna Products**
P.O. Box 297
Westfield, IN 46074

**W. S. Deans Co.**
8512 E. Gardendale Ave.
Downey, CA 90242

**Delta Manufacturing**
P.O. Box 27
Lorimor, IA 50149

**Electro-Craft Systems**
924 Fern Grove Drive
San Jose, CA 95129

**Elgas Racing Enterprises**
1235 S. 23rd St.
Phoenix, AZ 85014

**Euro Imports**
19H Gardner Rd.
Fairfield, NJ 07006

**Jerobee** (see JoMac)

**JoMac Products Inc.**
12702 N.E. 124th St.
Kirkland, WA 98033

**K & B Manufacturing**
12152 S. Woodruff Ave.
Downey, CA 90241

**KRD Products**
P.O. Box 3391
Shawnee, KS 66203

**LaTrax Corp.**
2714-B Kingsley
Garland, TX 75041

**Leisure Electronics**
11 Deerspring Rd.
Irvine, CA 92705

**Linex Racing Cars**
623 Olympic Blvd. Suite 17
Santa Monica, CA 90401

**Dick McCoy/C&H Inc.**
10767 Monte Vista Ave.
Ontario, CA 91761

**MACH-12 R/C Products**
4539 California Ave., S.W.
Seattle, WA 98116

**Marker Machine Inc.**
5240 N. 124th St.
Milwaukee, WI 53225

**Mintec Co.**
209 Alverno Drive
Fort Wayne, IN 46816

**Model Racing Products (MRP)**
12702 N.E. 124th St.
Kirkland, WA 98033

**Model Rectifier Corp. (MRC)**
2500 Woodbridge Ave.
Edison, NJ 08817

**Otaki** (see Scale Craft)

**Orbit Electronics**
1641 Kaiser Ave.
Santa Ana, CA 92708

**Parma International**
4651 W. 130th St.
Cleveland, OH 44135

**Peerless/Kyosho**
3919 M St.
Philadelphia, PA 19124

**Penn Racing Specialties**
4316 Clareville Drive
Allison Park, PA 15101

**Ridge Industries**
P.O. Box 482
Montclair, NJ 07040

**Scale Craft Models Inc.**
8735 Shirley Ave.
Northridge, CA 91324

**Tamiya** (see Model Rectifier)

**Tatone Products**
1209 Geneva Ave.
San Francisco, CA 94112

**Taurus** (see V.M.W.)

**Testor Corp.**
620 Buckbee St.
Rockford, IL 61101

**Thorp Manufacturing**
1655 E. Mission Blvd.
Pomona, CA 91766

**Thunder Road Automotive**
1209 Geneva Ave.
San Francisco, CA 94112

**Twinn-K Inc.**
P.O. Box 31228
Indianapolis, IN 46231

**Ultimara Inc.**
8670 Chardon Rd.
Kirtland, OH 44094

**Veco** (see K & B Manufacturing)

**V.M.W. Co.**
1640 E. Edinger Ave.
Santa Ana, CA 92705

**Workrite R/C Hobby**
7009 Beaty Ave.
Fort Wayne, IN 46809

## RADIO CONTROL EQUIPMENT

**Cannon Electronics**
13400-26 Saticoy St.
North Hollywood, CA 91605

**Cirrus/Hobby Shack**
18480 Bandilier Circle
Fountain Valley, CA 92708

**Cox/Sanwa**
1505 E. Warner Ave.
Santa Ana, CA 92702

**EK-logictrol**
3322 Stovall St.
Irving, TX 95061

**Futaba**
630 W. Carob St.
Compton, CA 90220

**Lewis Galoob, Inc.**
333 Alabama St.
San Francisco, CA 94110

**Heathkit**
Heath Company
Benton Harbor, MI 49022

**Kraft Systems Inc.**
450 W. California Ave.
Vista, CA 92083

**Litco Systems**
P.O. Box 90
East Hanover, NJ 07936

**Model Rectifier Corp. (MRC)**
2500 Woodbridge Ave.
Edison, NJ 08817

**Pro-cision Products**
6501 Flotilla St.
Los Angeles, CA 90040

**Royal Electronics Corp.**
3535 S. Irving St.
Englewood, CO 80110

**RS Systems**
5301 Holland Drive
Beltsville, MD 20705

**S&O R/C Products**
23700 Bessemer St.
Woodland Hills, CA 91367

**Specialist**
Millicott Corp.
1420 Village Way
Santa Ana, CA 92705

**Westport International Inc.**
349 Boston Post Rd.
Milford, CT 06460

## ASSEMBLY MATERIALS, TOOLS, ACCESSORIES

**Badger Air-Brush Co.**
9128 W. Belmont Ave.
Franklin Park, IL 60131

**Bammco**
Box 1334
Canoga Park, CA 91304

**CPC**
777 W. Grand Ave.
Oakland, CA 94612

**Custom Craft Products**
19 Florgate Rd.
Farmingdale, NY 11735

**Delta Manufacturing**
P.O. Box 27
Lorimor, IA 50149

**Dremel Manufacturing**
Division, Emerson Electric Co.
4915 21st St.
Racine, WI 53406

**Du Bro Products Inc.**
480 Bonner Rd.
Wauconda, IL 60084

**F.A.I. Model Supply**
1800 W. Hatcher Rd.
Phoenix, AZ 85021

**Floquil**
Route 30
North Amsterdam, NY 12010

**Fox Manufacturing Co.**
5305 Towson Ave.
Fort Smith, AR 72901

**M. Grumbacher Inc.**
460 W. 34th St.
New York, NY 10001

**Hobbypoxy** (see Pettit Paint)

**K & B Manufacturing**
12152 Woodruff Ave.
Downey, CA 90241

**K & S Engineering**
6917 W. 59th St.
Chicago, IL 60638

**Krasel Industries Inc.**
1821 E. Newport Circle
Santa Ana, CA 92705

**The K.J. Miller Corp.**
2401 Gardner Rd.
Broadview, IL 60153

**Model Builder Products**
621 W. 19th St.
Costa Mesa, CA 92627

**Paasche Air-Brush Co.**
1909 W. Diversey Parkway
Chicago, IL 60614

**Pactra Industries**
7060 Hollywood Blvd.
Los Angeles, CA 90028

**Penn Racing Specialties**
4316 Clareville Drive
Allison Park, PA 15101

**Pettit Paint Co. Inc.**
20 Pine St.
Rockway, NY 07866

**Scalecoat**
**Quality Craft Models, Inc.**
177 Wheatley Ave.
Northumberland, PA 17857

**Tatone Products**
1208 Geneva Ave.
San Francisco, CA 94112

**Testor Corp.**
620 Buckbee St.
Rockford, IL 61101

**Thorp Manufacturing**
1655 E. Mission Blvd.
Pomona, CA 91766

**Top Flite**
1901 N. Narragansett Ave.
Chicago, IL 60639

**Twinn-K Inc.**
10296 W. Washington St.
P.O. Box 31228
Indianapolis, IN 46231

**V.M.W. Co.**
1640 E. Edinger Ave.
Santa Ana, CA 92705

**Wen Products**
5810 Northwest Hwy.
Chicago, IL 60631

**X-Acto Inc.**
45-35 Van Dam St.
Long Island City, NY 11101

# Advice from the Experts

ACTIVE HOBBYISTS always seem to be looking for tips that will enable them to increase their skill. The result is that each new model car they assemble is better than the one before. And fortunately, experienced modelers are eager to pass along some of the procedures they have picked up during their pursuit of the hobby.

Many shortcuts are the result of a bit of brainstorming on the part of modelers who are forced to make do with less equipment than they would like to have. Other tips come as modelers learn from their mistakes. A tip in this section that could have easily resulted from a mistake is the one about the use of lacquer on a plastic model. The chemical reaction between the paint and the plastic, while unacceptable for the car's body, gives a desirable look to the car seats.

In the following pages, some experts present hints intended to help other modelers produce cars that look very realistic and run smoothly.

Plastic parts do not always fit together perfectly without alteration. Before you begin cementing, fit the parts together to see that the surfaces mate precisely. If the parts fit together badly, use a hobby knife and fine-grit sandpaper to change the shape of the parts. Trim only a small amount of plastic at a time, checking for good fit after each trimming.

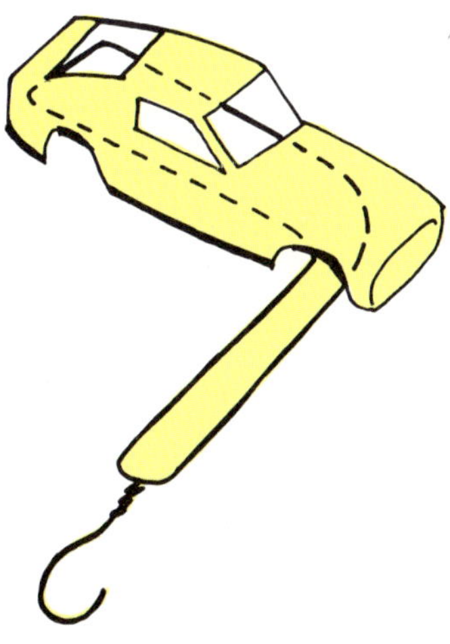

An ordinary coat hanger makes an excellent spraying stand when bent to fit the shape of the car's underside. After painting, place the model in a shoe box to protect it from dust while drying. Wait three or four hours before handling the car.

To make plastic model car seats look like they're upholstered with vinyl, spray them with lacquer. The lacquer attacks the plastic, causing what is known as crazing. The result is a wrinkled finish that gives the seat the appearance of vinyl upholstery.

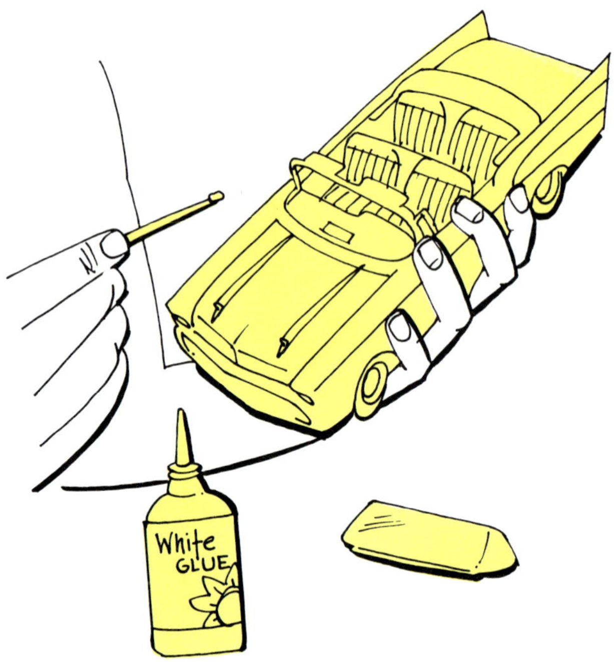


Clear windshields should be glued in place after all other assembly steps are complete. Hold the windshield in place and use a toothpick to apply a small amount of white glue such as Elmer's Glue-All to each corner. The glue cannot harm the plastic and will dry clear.

Adding fuel lines and ignition wires to a plastic model car engine will give it a more realistic appearance. Use red, black or yellow thread for ignition wires and clear fishing line for fuel lines.

Holes for steering wheel assemblies and headlights that are of the incorrect depth or are filled in with excess metal can be redrilled. Never use a power drill, especially on soft metal kits; use a small drill bit in a pin chuck and finger pressure only to avoid enlarging the hole too much.

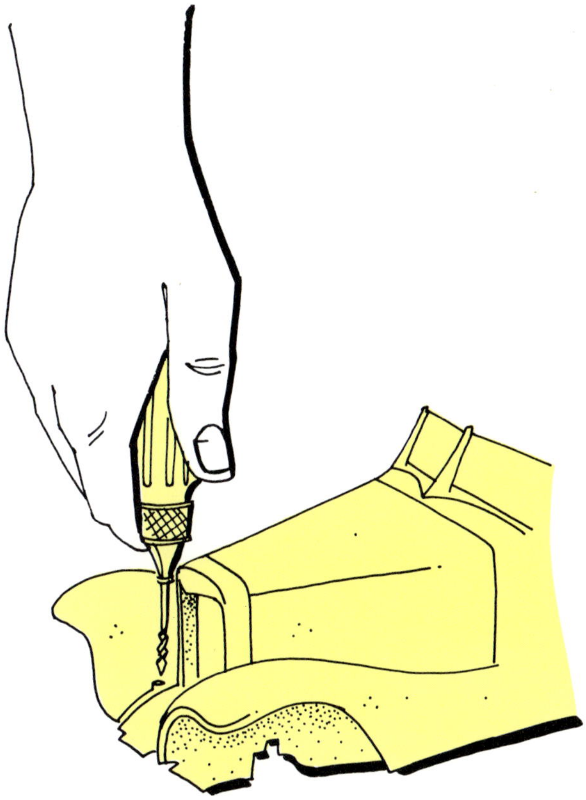

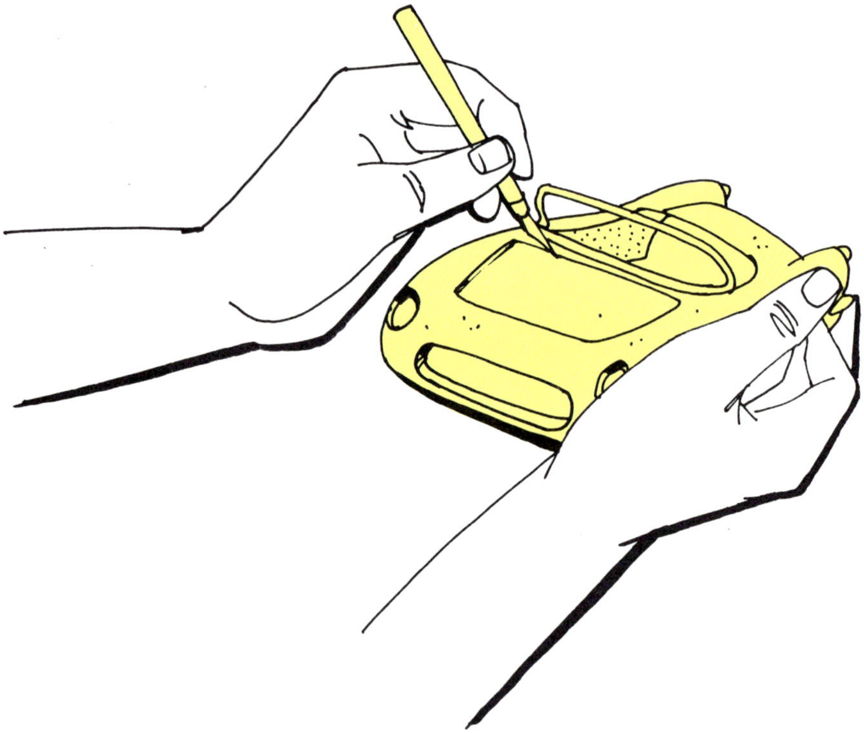

Examine the metal model's parts before actually assemblying the car. If you find flash marks along mold lines, scrape them smooth with a hobby knife. A small file can also be used.

If you find that the casting is uneven on your metal kit car, you can apply ordinary automotive repair putty and sand it down to a smooth finish. The putty can be found in auto parts stores.

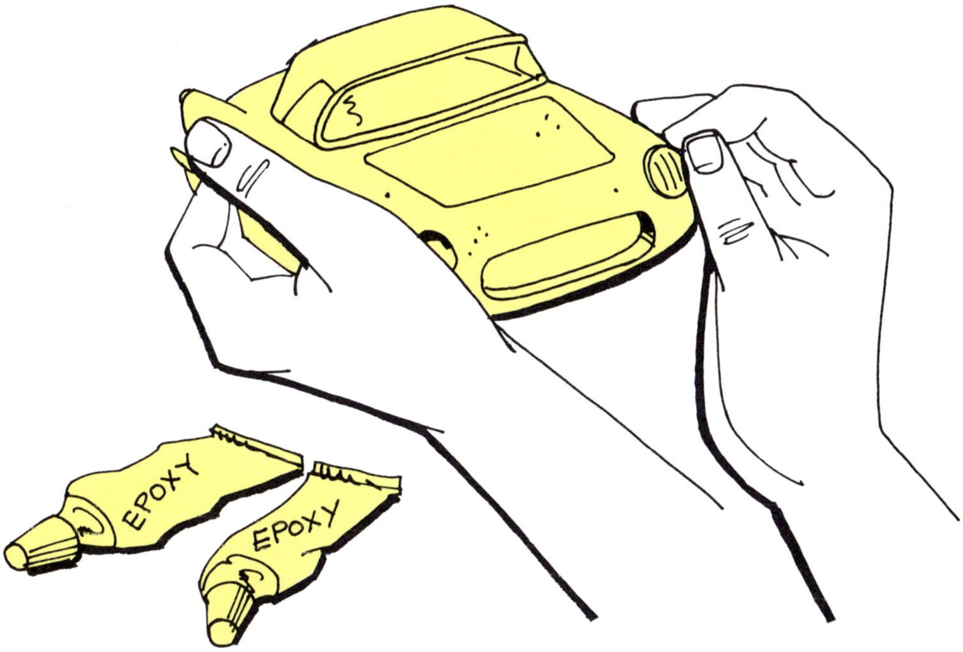


Five-minute epoxy-type glue should be used to cement styrene plastic parts to metal parts of the car. Mix the two components of the epoxy, and apply a small amount to the location where the plastic part is to be secured. Hold the part in place until the epoxy sets. White glue can be used as a substitute for epoxy, but it will not work as well. If you use white glue, first scrape the surfaces of the two mating parts with a hobby knife to increase adhesion.

If plastic decals will not stick to a metal model, do not use glue as a cement to affix them to the body. Instead, use clear varnish.

You can improve your slot car's performance by altering the stock shape of its pickup shoes. Use tweezers or long-nose pliers to bend the shoes so that they have optimum contact with the rails.

Run sanding your slot car's rear tires can lower the chassis and round the edges of the tires slightly to reduce flipping. Cement a piece of fine sandpaper to a thin piece of wood or cardboard. Then place your car on the track, rev the car's engine to full speed and touch its tires to the sandpaper. Be careful not to sand too much, because the reduced tire diameter will cause the chassis to scrape the track. Clean the tires by running them over the sticky side of masking tape.

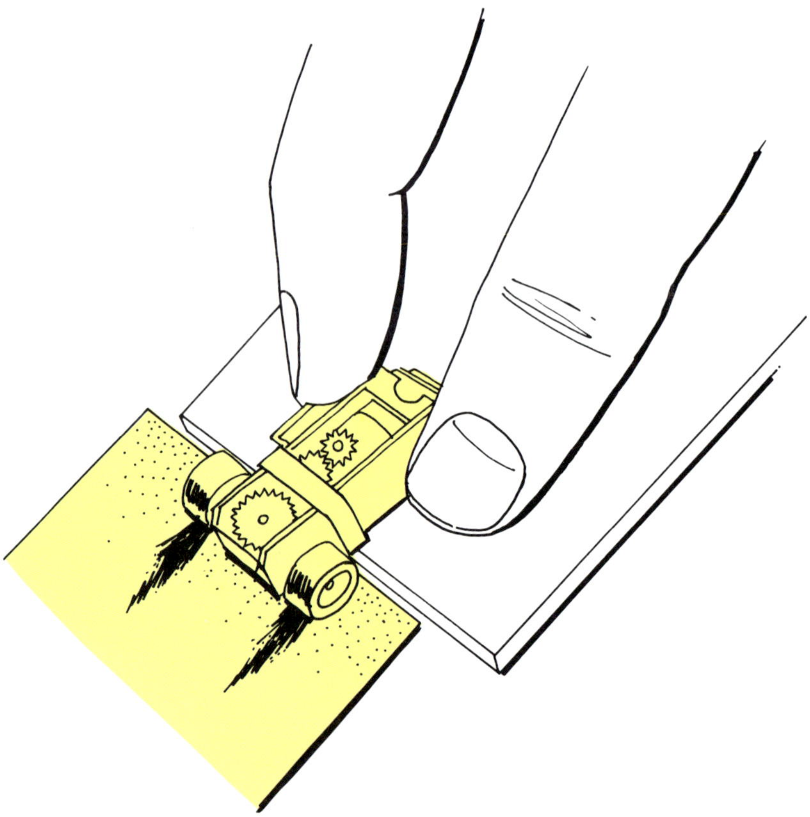

Carbon compound motor brushes of slot cars can be cleaned by rubbing them on a sheet of clean paper. If you find that the brushes are worn, replace them. Worn brushes rob your motor of power.

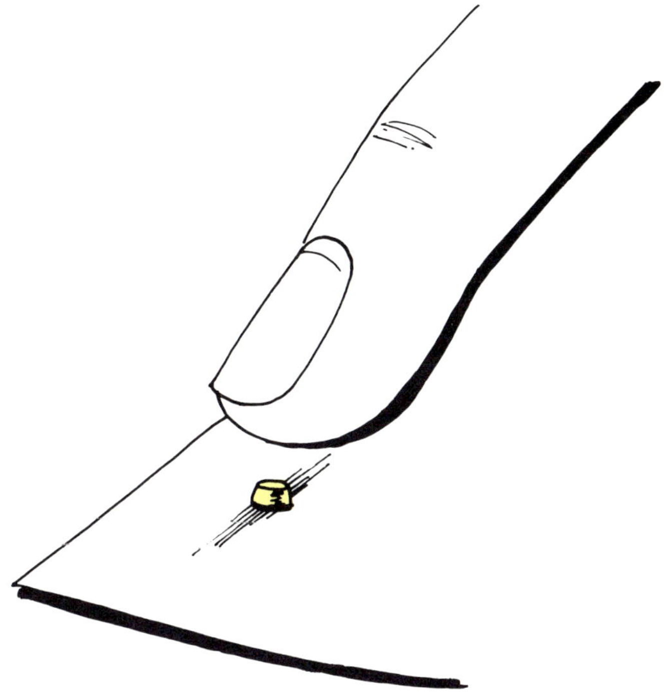

Uneven track rails will cause a slot car to lose performance. Use a feeler gauge to determine whether they are positioned at the recommended height. Level uneven rails with a small block of wood and a hammer. Place the block on the rails and tap it with a hammer until the rails are even.

Rails that are too low can be pushed up from below with a screwdriver. Mounting fuel tanks too low in a radio control car forces the engine to pull fuel uphill and can cause problems. For this reason, it is best to mount the tank level with the battery box on the car's radio receiver.

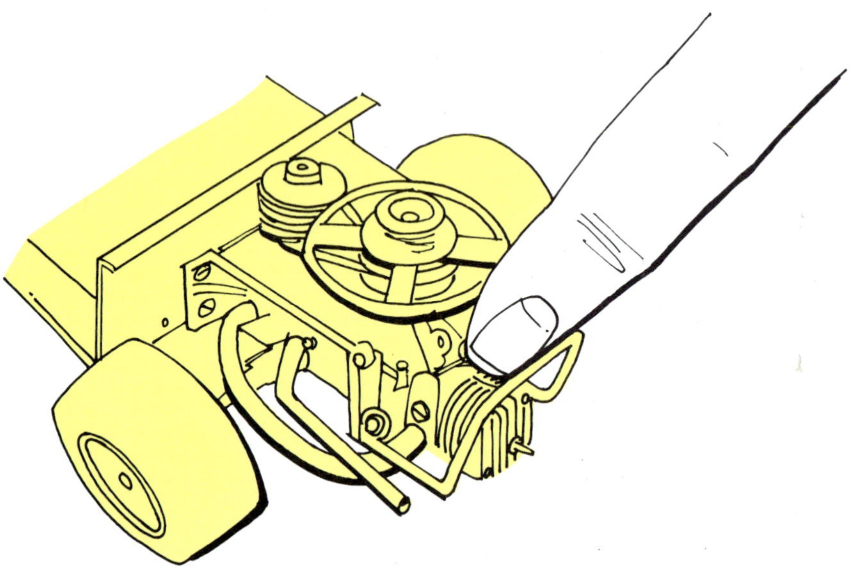

If your RC car's engine is hard to start when cold, the problem may be a fuel/air mixture that is too lean. Try choking the engine as it turns over by placing your finger over the screen-covered air intake at the rear of the engine.

An RC car engine will heat up faster when waiting at the starting line than when the car is in motion. Dripping fuel on the engine will cool it, and help prevent the power plant from quitting.

If your RC car spins out when turning in one direction but handles well when turning in the opposite direction, adjust the front suspension to put more weight on the inside rear tire.